Wireless Networks

Series editor

Xuemin (Sherman) Shen
University of Waterloo, Waterloo, Ontario, Canada

More information about this series at http://www.springer.com/series/14180

Tho Le-Ngoc • Ahmed Masmoudi

Full-Duplex Wireless Communications Systems

Self-Interference Cancellation

Springer

Tho Le-Ngoc
Department of Electrical and Computer Engineering
McGill University
Montréal, QC, Canada

Ahmed Masmoudi
Department of Electrical and Computer Engineering
McGill University
Montréal, QC, Canada

ISSN 2366-1186 ISSN 2366-1445 (electronic)
Wireless Networks
ISBN 978-3-319-57689-3 ISBN 978-3-319-57690-9 (eBook)
DOI 10.1007/978-3-319-57690-9

Library of Congress Control Number: 2017943056

Printed on acid-free paper

This Springer imprint is published by Springer Nature
The registered company is Springer International Publishing AG
The registered company address is: Gewerbestrasse 11, 6330 Cham, Switzerland

Preface

Full-duplex operation for wireless communications can potentially double the spectral efficiency, compared to half-duplex operation, by using the same wireless resource to transmit and receive at the cost of a large power difference between the high-power self-interference (SI) from its own transmitted signal and the low-power intended signal received from the other distant transceiver. The SI can be gradually reduced by a combination of radio-frequency (RF) and baseband cancellation stages. Each stage requires the estimation of the different distortions that the SI endures such as the SI channel and the transceiver nonlinearities. This book deals with the development of SI-cancellation techniques that are well adapted to the full-duplex operation.

First, we recognize the sparseness of the SI channel and exploit it to develop a compressed-sensing (CS) based SI channel estimator. The obtained estimate is used to reduce the SI at the RF prior to the receiver low-noise amplifier and analog-to-digital converter to avoid overloading them. To further reduce the SI, a subspace-based algorithm is developed to jointly estimate the residual SI channel, the intended channel between the two transceivers, and the transmitter nonlinearities for the baseband cancellation stage. Including the unknown received intended signal in the estimation process represents the main advantage of the proposed algorithm compared to previous data-aided estimators that assume the intended signal as additive noise. By using the second-order statistics of the received signal, it is possible to obtain the noise subspace and then to estimate the different coefficients without knowing the intended signal. Depending on the number of transmit and receive antennas, we propose to use either the received signal or a combination of the received signal and its complex conjugate. Also, we develop a semi-blind maximum likelihood (ML) estimator that combines the known pilot and unknown data symbols from the intended transceiver to formulate the likelihood function. A closed-form expression of the ML solution is first derived, and an iterative procedure is developed to further improve the estimation performance at moderate to high signal-to-noise ratio. Simulations show significant improvement in SI-cancellation gain compared to the data-aided estimators.

Moreover, we present two new SI-cancellation methods using active signal injection (ASI) for full-duplex MIMO-OFDM systems. The ASI approach adds an appropriate cancelling signal to each transmitted signal such that the combined signals from the transmitting antennas attenuate the SI at the receive antennas. In the first method, the SI-pre-cancelling signal uses some reserved subcarriers which do not carry data. In the second method, the constellation points are dynamically extended within the constellation boundary in order to minimize the received SI. Thus, the SI-pre-cancelling signal does not affect the data-bearing signal. Simulation results show that the proposed methods considerably reduce the SI at a modest computational complexity.

This book aims to provide researchers and engineers in wireless communications, an understanding of the challenges and practical solutions to implement a full-duplex system.

This work was partially supported by the Natural Sciences and Engineering Research Council (NSERC) and Huawei Technologies Canada.

Montréal, QC, Canada
November 2016

Tho Le-Ngoc
Ahmed Masmoudi

Contents

Acronyms

ADC	Analog-to-digital converter
AS	Antenna selection
ASI	Active signal injection
BER	Bit error rate
BS	Beam selection
CRB	Cramér-Rao bound
CS	Compressed sensing
DAC	Digital-to-analog converter
dB	Decibel
dBm	Decibel-milliwatts
DoF	Degree of freedom
FFT	Fast Fourier transform
IFFT	Inverse fast Fourier transform
IIP	Input intercept point
IQ mixer	In-phase/quadrature mixer
LNA	Low noise amplifier
LoS	Line-of-sight
LS	Least square
MAC	Media access control
MIMO	Multiple-input multiple-output
ML	Maximum likelihood
MMSE	Minimum mean square error
MSE	Mean square error
OFDM	Orthogonal frequency-division multiplexing
OIP	Output intercept point
PA	Power amplifier
PLL	Phase-locked loop
PSK	Phase-shift keying
QAM	Quadrature amplitude modulation
RF	Radio-frequency
RIP	Restricted isometry property

SI	Self-interference
SINR	Signal-to-self-interference noise ratio
SIR	Signal-to-self-interference ratio
SISO	Single-input single-output
SNR	Signal-to-noise ratio
SVD	Singular value decomposition
VGA	Variable gain amplifier

Chapter 1
Introduction

Abstract Full-duplex operation of wireless communications can potentially double the spectral efficiency, compared to half-duplex operation, by using the same wireless resource simultaneous transmission and reception at the costs of a large power difference between the high-power self-interference (SI) from its own transmitted signal and the low-power intended signal received from the other distant transceiver. The SI can be gradually reduced by a combination of radio-frequency (RF) and baseband cancellation stages. Each stage requires the estimation of the different distortions that the SI endures such as the SI channel and the transceiver nonlinearities. This chapter presents the idea of each cancellation stage and the potential applications of full-duplex communications.

1.1 Full-Duplex Wireless Systems

Traditionally, wireless communications systems operate in half-duplex mode where a transceiver transmits and receives in non-overlapping time-slots, i.e., time-division duplex (TDD), or frequency-slots, i.e., frequency-division duplex (FDD), or in different orthogonal spectrum-spreading codes, to avoid the possible strong self-interference (SI) from its own transmission to its reception. As a much higher spectral efficiency is required to support the fast growth of wireless communications applications, full-duplex operation by simultaneous transmission and reception over the same frequency-slot is an attractive solution to potentially double the spectral efficiency if the resulting SI can be cancelled or suppressed to a sufficiently low level for proper detection of the low-power intended received signal from the other transceiver.

Full-duplex operation is not new and has been successfully used in wireline communications for a long time. Here, the interference, also called line echo, results from the coupling between the transmit and the receive wires and from the impedance mismatch when converting the 4-wire interface to the 2-wire interface through the hybrid, as illustrated in Fig. 1.1. This line echo is 3–6 dB lower than the intended signal [1, 2] making the required cancellation level relatively low in the range of 20–30 dB.

One may wonder why current wireless communications systems do not operate in full-duplex mode. When the transceivers communicate in a full-duplex fashion,

T. Le-Ngoc, A. Masmoudi, *Full-Duplex Wireless Communications Systems*, Wireless Networks, DOI 10.1007/978-3-319-57690-9_1

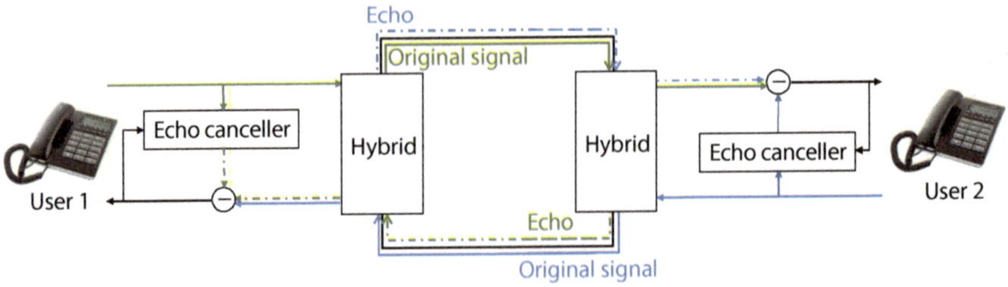

Fig. 1.1 Echo cancellation in a wireline communication system

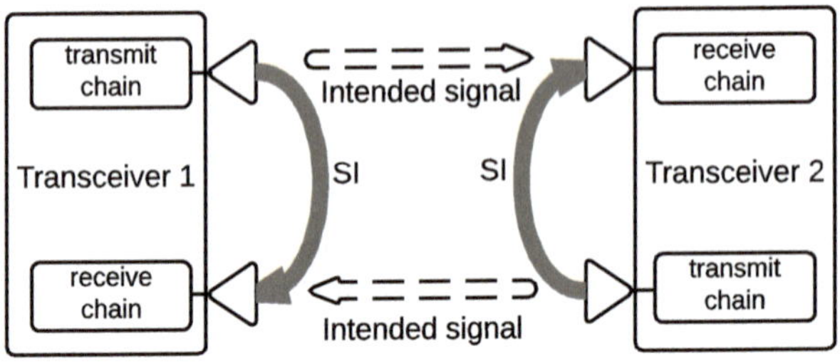

Fig. 1.2 SI in a full-duplex point-to-point wireless communication system

the receiver experiences co-channel SI from its own transmitter, as shown in Fig. 1.2. This SI is usually several orders of magnitude higher than the intended signal because the latter crosses longer distance than the SI. For example, considering two transceivers distant by 500 m, the intended signal coming from the distant transceiver is attenuated by approximately 120 dB. If there is 15 dB isolation between the transmit and the receive paths of the same transceiver, then the SI would be 105 dB higher than the intended signal. This huge difference between the power levels of the SI and the intended signal increases with more distance between the two communications transceivers. Therefore, a much higher SI-cancellation is required in full-duplex *wireless* systems than echo cancellation in full-duplex *wireline* communications. This power difference dictates the choices of the SI-cancellation techniques and strategies to achieve the challenging high-SI-cancellation requirements.

Given that the transmitted SI is known, it could be used to remove the SI from the received signal. If this operation is done in the digital domain at baseband, after the analog-to-digital converter (ADC), then the ADC dynamic range will represent a major bottleneck. Actually, the input to the ADC is scaled so that the level of the strong SI matches the dynamic range of the ADC. According to the classical rule of thumb for a 10 bit ADC, the resulting quantization noise is $6.02 \times 10 + 1.76 = 61.96$ dB lower than the signal at the input of the ADC. If the SI is 100 dB higher than the intended signal, then the quantization noise will be about 38 dB higher than the intended signal. Therefore, even if the SI is completely cancelled at the output of the ADC, the receiver is no longer able to process the intended signal. To

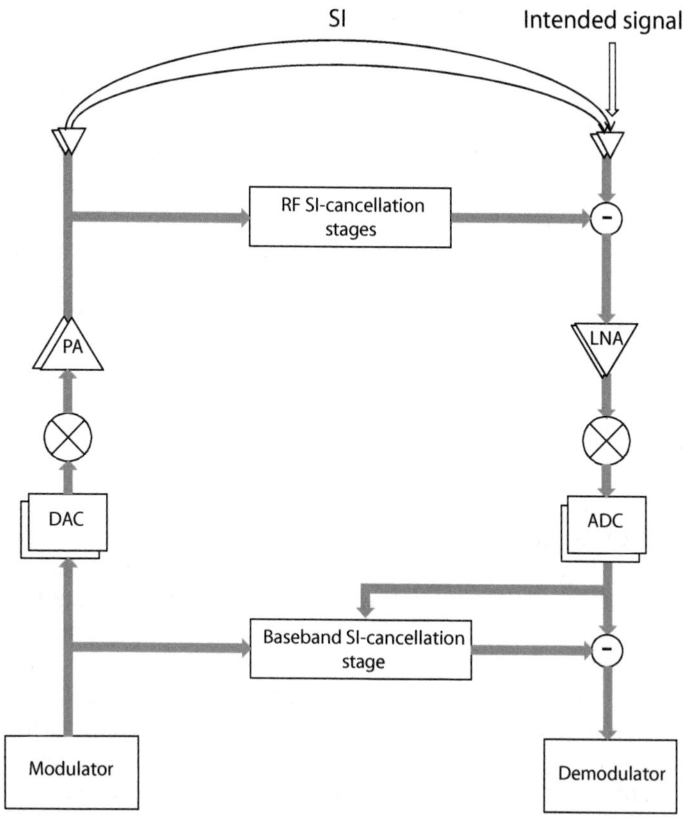

Fig. 1.3 Simplified block diagram of a full-duplex transceiver with RF and baseband SI-cancellation stages

avoid this problem, the SI must be first reduced at the input of the receiver prior to the low-noise amplifier (LNA) and the ADC, as shown in Fig. 1.3. Then, the SI can be further reduced at the baseband after the ADC to improve the detection of the intended signal. Moreover, the transmitter impairments are of significant level compared to the received intended signal and need also to be reduced [3, 4]. Thus, simply reducing the SI based on the known transmitted symbols can result in a large residual SI.

1.2 SI-Cancellation Techniques

Recently, a large variety of SI-cancellation techniques have been proposed for full-duplex systems [4–7]. The proposed approaches use a combination of antenna techniques, radio-frequency (RF) techniques and baseband techniques.

1.2.1 Antenna SI-Cancellation Techniques

The antenna SI-cancellation techniques aim to reduce the SI impinging upon the receiving antennas by a proper design of the transmit and the receive antenna structures. Antenna SI-cancellation can be achieved by using antenna separation, polarization, and isolation [8–11], directional antennas [12–14] or antennas placement to create null space at the receive antennas [15, 16]. The applicability of each one of these methods depends on the application and the physical constraints of the system. For example, in mobile applications with small device dimensions, the passive suppression achieved using antenna separation and isolation is very limited. However, in other systems (e.g., relay systems) where the transmit and receive antennas are not necessary collocated, antenna separation and isolation could achieve a significant amount of reduction. For instance, in [10], the use of a directional antenna and 4–6 m of antenna separation achieves about 80 dB of suppression. This large antenna separation might be acceptable in relay systems, but it is not acceptable in practical mobile applications. A more practical passive self-interference suppression method with relatively small antenna separation (i.e., 20–40 cm) is introduced in [17]. The results show a maximum of 60 dB passive suppression with cross polarization, and a metal shield between the antennas. A reconfigurable antenna is proposed in [18] where the main beam direction of the antenna can be directed into a desired direction by the proper reactive loading of the parasitic elements to maximize the signal-to-SI-and-noise power ratio.

When building a full-duplex transceiver, we have the choice between two methods of interfacing antennas. Either we use physically separate antennas for the transmission or the reception, or we use one antenna to simultaneously transmit and receive, where the transmission and reception paths are isolated through a circulator. In the separate antenna architecture, as illustrated in Fig. 1.4, the internal reflections comprise the signal propagating directly from the transceiver's transmit antennas to its receive antennas and the reflection from the transceiver structure. Whereas in the shared antenna architecture, the internal reflections come from the antenna impedance mismatch and the circulator leakage that does not perfectly isolate the transmit and receive paths, (e.g., a typical transmit-receive isolation of about 20 dB from commercially available circulators). These internal reflections are static since they depend on the structure of the transceiver. The external reflections create multiple copies of the SI, which may vary over time and thus limit the isolation provided by the antenna designs.

1.2.2 RF SI-Cancellation Techniques

The RF SI-cancellation aims to suppress the SI before the LNA and ADC by subtracting an estimate of the received SI from the received signal. Figure 1.3 shows the general structure of the transceiver where the RF transmitted signal

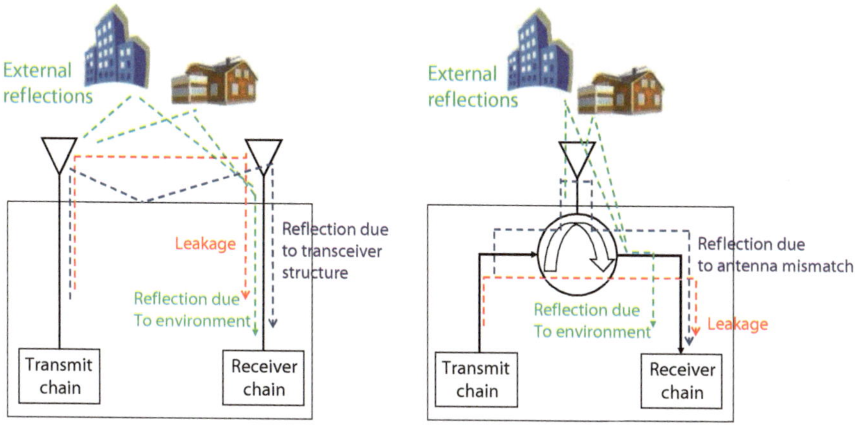

Fig. 1.4 Illustration of the SI channel for separate-antenna and shared-antenna architectures

can be extracted at the transmit PA output, processed in the RF SI-cancellation stage and subtracted from the received signal. Analog RF SI-cancellation can be first applied [13, 14, 17] to suppress only the internal coupling and reflections modelled by a programmable analog tapped-delay line (TDL) transversal filter. Further adaptive digital RF cancellation can also be applied to suppress the SI components coming from the random external reflections by using a digital symbol-synchronous finite-impulse-response (FIR) filter [5, 6, 12, 19, 20]. The analog and digital RF SI-cancellation stages will be discussed in detail in Chap. 2.

1.2.3 Baseband SI-Cancellation Techniques

The baseband SI-cancellation aims to reduce the residual SI after the ADC by applying various signal processing techniques to the received signal [7, 21]. For the baseband SI-cancellation to be successful, the SI should be sufficiently reduced before the ADC, using the antenna and RF SI-cancellation stages. The advantage of working in the digital domain is that advanced digital processing becomes relatively easy to perform. In order to subtract the received SI, we need to capture every modification that may happen to the transmitted SI including the propagation channel and the nonlinearities of the RF components such as the in-phase/quadrature (IQ) mixer and the power amplifier (PA). This requires an effective estimation of the SI channel, and the transceiver impairments in order to create an accurate replica of the received SI signal.

Furthermore, spatial domain cancellation attempts to reduce the SI by precoding at the transmit chain and postcoding at the receive chain. More specifically, precoding and postcoding modify the transmission to reduce the SI. They utilize the degrees of freedom (DoF) provided by the multiple-input multiple-output (MIMO) systems, by confining the transmit and receive signals to a subset of the available space. This operation sacrifices some available antennas for SI-cancellation. For example, it was reported in [22] that a physical 4×4 MIMO is used as a 2×2 MIMO for data transmission and the rest of the antennas are used for SI-cancellation. This raises the question: is it possible to perform spatial cancellation without affecting the DoF provided by MIMO systems? Moreover, these techniques suppose the knowledge of the SI and intended channels which motivates the development of channel estimators for full-duplex systems even more. Other techniques transmit a cancelling signal with the data-bearing signal to reduce the received SI [23].

In the next chapters, we discuss the existing techniques in more details, improve some of them and propose new cancellation techniques.

1.3 Applications of Full-Duplex Communications

In addition to point-to-point communications, Fig. 1.5 presents other basic topologies as potential candidates to work in a full-duplex fashion. First, consider a full-duplex base-station sending data on the downlink to one half-duplex user and

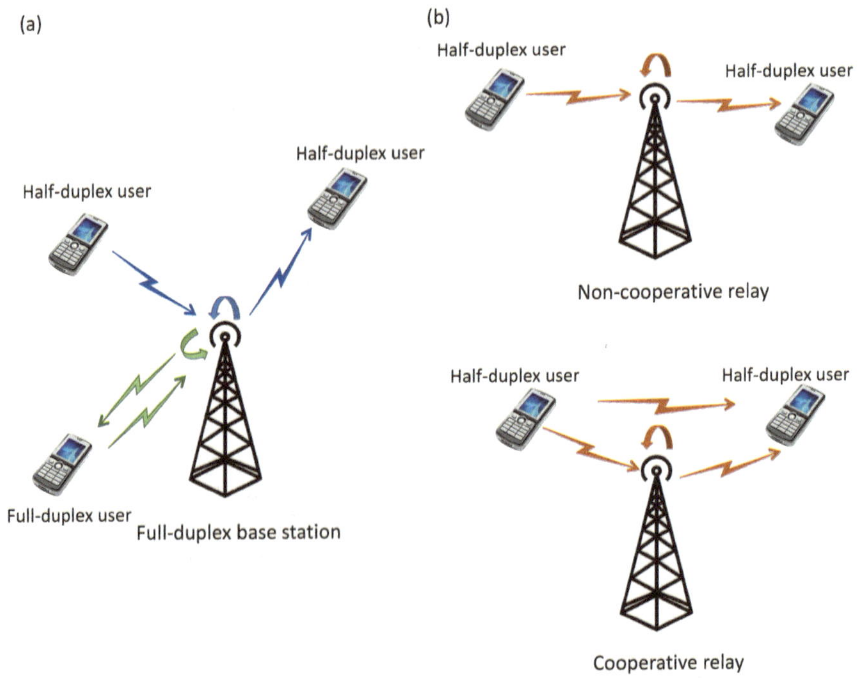

Fig. 1.5 Examples of topologies operating in a full-duplex fashion. (**a**) Point-to-multipoint. (**b**) Relaying

receiving data on the uplink from another half-duplex user. In this case, the base station can send and receive simultaneously over the same frequency slot while, in half-duplex mode, it would need two time/frequency resources to transmit and receive. This requires sufficient separation between the two users to avoid the interference caused by the uplink user on the downlink user. In the presence of a full-duplex user, the uplink and downlink with that user can be performed over the same time/frequency resource. Furthermore, the frequency planning can get simpler as a single frequency is needed for both the uplink and the downlink. Second, a full-duplex relay station receives and forwards simultaneously the signal between two half-duplex terminals. Thus, the relay can increase the spectral efficiency compared to half-duplex operation.

The past few years have witnessed a growing deployment of small cell sizes due to the prominence of WiFi, femtocell, picocell, etc. The short distance between the communicating nodes reduces the transmit and receive power difference and hence the required SI-cancellation level, which increases the interest in full-duplex operation. To understand this relation, we consider the following simple example. To serve one user at the edge of the cell, a base station has to increase its transmit power to compensate the pathloss. In large cells, the transmit power is higher compared to smaller cell to achieve the same signal-to-noise ratio. Therefore, the transmitter observes larger receive power difference between the intended signal and the SI compared to small cells. For small cells, the received SI can be managed, making full-duplex transmission easier.

In addition to spectral efficiency improvement, full-duplex communications can also improve the overall throughput of a wireless network. It removes the hidden terminal problem and the resulting collision and retransmission. To understand this point, we consider three half-duplex nodes A, B and C communicating. We suppose that nodes A and C cannot hear each other and they send packets simultaneously to node B. This means that the packets will collide and need to be retransmitted. If now the nodes use full-duplex operation, node B will also transmit to node A at the same time and in the same frequency slot as node A to node B. Thus, node C is able to sense the medium as busy and hence moves its transmission to another time-slot or frequency-slot. This feature opens new media access control (MAC) layer protocols that use the simultaneous transmission and reception [6, 12].

In a cognitive radio network, full-duplex transmission allows the secondary terminals to sense the traffic in the network during their own transmission [24]. They do not need to stop transmitting in order to listen to the channel and they can immediately stop transmitting when the primary terminal starts using the channel. The feasibility of this application requires the power of the residual SI, after cancellation, to be lower than the power of the received primary user.

Full-duplex transmission can also be applied to improve the security of wireless data transfer [25, 26]. Here, the receiver transmits a jamming signal simultaneously while receiving in such a way that the eavesdropper receives a superposition of the useful signal and the jamming signal. Without any prior knowledge of the structure of the two signals, it is difficult for the eavesdropper to detect the useful signal.

1.4 Book Organization

The primary objective of this book is to discuss new efficient SI-cancellation techniques for full-duplex wireless communications.

In Chap. 2, we start with a brief survey of the most relevant state-of-the-art cancellation techniques. We summarize the existing architectures to reconstruct and cancel the SI. Then, a brief overview of the existing estimation algorithms used to reconstruct the SI for SI-cancellation is presented. We also summarize the spatial cancellation techniques applicable to SI suppression. This background material will be the starting point for the developments of new algorithms and methods for SI-cancellation.

Various experimental results have indicated that the SI in full-duplex communications can be mitigated to properly detect the intended signal. However, relatively little is known about the cancellation limits of these systems. In Chap. 3, we investigate the basic SI-cancellation bottlenecks in full-duplex wireless systems. To that end, we first classify the known full-duplex architectures based on where the reference signal is taken from to cancel the SI. By combining the effects of the transceiver impairments, the estimation error and the SI channel, our analysis reveals that the main bottleneck to completely cancel the SI turns out to be either the quantization noise, the phase noise in the local oscillator or the estimation error, depending on the used architecture. We provide comprehensive numerical results that justify the need of applying RF cancellation stage and reducing the transmitter impairments in the baseband cancellation stage.

Next, we turn our attention to the developed estimation algorithms. By exploiting the SI channel sparsity, a compressed-sensing (CS) based SI channel estimation technique presented in Chap. 4 is used in the digital RF SI-cancellation stage to reduce the SI power prior to the receiver LNA and the ADC to avoid overloading them. The sensing matrix is proved to satisfy the restricted isometry property (RIP). Subsequently, a subspace-based algorithm is proposed to jointly estimate the coefficients of both the residual SI and intended channels, and the transceiver impairments for the baseband SI-cancellation stage to further reduce the residual SI. The objective is to develop an algorithm that estimates the residual SI in the presence of the unknown intended signal. Therefore, we include the intended signal in the estimation process instead of considering it as additive noise. It is demonstrated that the SI channel coefficients can be accurately estimated without any knowledge of the intended signal, and only few training symbols are needed for ambiguity removal in intended channel estimation. By comparing the mean square error (MSE) performance of the proposed algorithm with that of the data-aided estimator, we demonstrate that the subspace-based algorithm offers superior estimation accuracy and higher SI-cancellation.

The subspace method proposed in Chap. 4 relies on the orthogonality property between the signal and noise subspaces. These two subspaces are obtained from eigen-decomposition of the covariance matrix of the received signal. Since the received signal consists of the SI and intended signals, the dimension of the signal subspace in full-duplex operation is at least twice than in traditional half-duplex operation. Therefore, to make the subspace technique works in full-duplex context, the number of receive antennas should be twice the number of transmit antennas. For this reason, Chap. 5 focuses on developing a subspace-based algorithm that is suitable for full-duplex systems when symmetric links are assumed. Here, we exploit both the covariance and pseudo-covariance matrices of the received signal to effectively increase the dimension of the observation space while keeping the dimension of the signal subspace unchanged. An iterative procedure is developed to estimate the ambiguity term and decode the intended signal. Finally, we use simulation to show that the proposed subspace approach provides significant improvements in cancellation performance and bit error rate (BER) over the conventional data-aided approach.

Chapter 6 describes a proposed maximum-likelihood (ML) approach to jointly estimate the SI channel, the transmitter impairments and intended channel by exploiting its own known transmitted symbols and both the known pilot and unknown data symbols from the other intended transceiver. The ML solution is obtained by maximizing the likelihood function under the assumption of Gaussian received symbols. A closed-form solution is first derived, and subsequently an iterative procedure is developed to further improve the estimation performance at moderate to high signal-to-noise ratio (SNR). We establish the initial condition to guarantee the convergence of the iterative algorithm. In the presence of considerable phase noise from the oscillators, a phase noise estimation method is proposed and combined with the ML channel estimator to mitigate the effects of the phase noise. Illustrative results show that the proposed methods offer good cancellation performance close to the Cramér-Rao bound (CRB).

Chapters 4–6, as well as most of the existing works, reduce the SI by subtracting it from the received signal. In Chap. 7, we present a new SI-cancellation method that reduces the SI prior to the receiver LNA and ADC for full-duplex MIMO OFDM systems. The basic idea includes adding an appropriate pre-cancelling signal to the transmitted signal such that, by its effects, the SI is greatly reduced at the receiver input. The proposed structure allows for various methods to be developed. One important required property of the cancelling signal is that it should not affect the detection process at the other intended receiver. To that end, two methods are proposed in this chapter. In the first method, the cancelling signal has frequency support on some reserved subcarriers which are not used for data transmission. In the second method, the constellation points are dynamically extended within the cancellation boundary in order to minimize the received SI. A combination of both methods is also proposed to enhance the cancellation capability. The proposed techniques are simple to implement and do not require any change in the receiver structure.

Finally, Chap. 8 concludes this book.

References

1. T. Gänsler and G. Salomonsson, "Nonintrusive measurements of the telephone channel," *IEEE Trans. on Comm.*, vol. 47, no. 1, pp. 158–167, 1999.
2. M. Ho, J. M. Cioffi, and J. A. Bingham, "Discrete multitone echo cancelation," *IEEE Trans. on Comm.*, vol. 44, no. 7, pp. 817–825, 1996.
3. A. Masmoudi and T. Le-Ngoc, "Self-interference cancellation limits in full-duplex communication systems," in *Proc. IEEE Global Telecommun. Conf.*, Washington DC, USA, Dec. 2016.
4. D. Bharadia, E. McMilin, and S. Katti, "Full duplex radios," in *Proc. of the ACM SIGCOMM 2013*, Hong Kong, China, Aug. 2013.
5. M. Duarte, C. Dick, and A. Sabharwal, "Experiment-driven characterization of full-duplex wireless systems," *IEEE Trans. Wireless Comm.*, vol. 11, no. 12, pp. 4296–4307, 2012.
6. J. I. Choi, M. Jain, K. Srinivasan, P. Levis, and S. Katti, "Achieving single channel, full duplex wireless communication," in *Proc. ACM MobiCom*, Chicago, Illinois, USA, Sept. 2010.
7. A. Masmoudi and T. Le-Ngoc, "Channel estimation and self-interference cancellation in full-duplex communication systems," *IEEE Trans. Vehicular Technology*, vol. 66, no. 1, pp. 321–334, Jan. 2017.
8. A. Sahai, G. Patel, and A. Sabharwal, "Pushing the limits of full-duplex: Design and real-time implementation," *arXiv preprint arXiv:1107.0607*, 2011.
9. ——, "Asynchronous full-duplex wireless," in *Communication Systems and Networks (COMSNETS)*, Bangalore, India, Jan. 2012.
10. C. R. Anderson, S. Krishnamoorthy, C. G. Ranson, T. J. Lemon, W. G. Newhall, T. Kummetz, and J. H. Reed, "Antenna isolation, wideband multipath propagation measurements, and interference mitigation for on-frequency repeaters," in *Proc. IEEE SoutheastCon*, Greensboro, North Carolina, USA, Mar. 2004.
11. K. Haneda, E. Kahra, S. Wyne, C. Icheln, and P. Vainikainen, "Measurement of loop-back interference channels for outdoor-to-indoor full-duplex radio relays," in *Proc. of the Fourth European Conference on Antennas and Propagation (EuCAP)*, Barcelona, Spain, Apr. 2010.
12. M. Jain, J. I. Choi, T. Kim, D. Bharadia, S. Seth, K. Srinivasan, P. Levis, S. Katti, and P. Sinha, "Practical, real-time, full duplex wireless," in *Proc. of ACM Mobicom*, Las Vegas, Nevada, USA, Sept. 2011.
13. E. Everett, M. Duarte, C. Dick, and A. Sabharwal, "Empowering full-duplex wireless communication by exploiting directional diversity," in *Proc. ASILOMAR Signals, Syst., Comput.*, Pacific Frove, California, USA, Nov. 2011.
14. E. Everett, A. Sahai, and A. Sabharwal, "Passive self-interference suppression for full-duplex infrastructure nodes," *IEEE Trans. Wireless Comm.*, vol. 13, no. 2, pp. 680–694, 2014.
15. M. A. Khojastepour, K. Sundaresan, S. Rangarajan, X. Zhang, and S. Barghi, "The case for antenna cancellation for scalable full-duplex wireless communications," in *Proc. of the 10th ACM Workshop on Hot Topics in Networks*, Cambridge, Massachusetts, USA, Nov. 2011.
16. J. I. Choi, S.-K. Hong, M. Jain, S. Katti, P. Levis, and J. Mehlman, "Beyond full duplex wireless," in *Proc. ASILOMAR Signals, Syst., Comput.*, Pacific Frove, California, USA, Nov. 2012.
17. M. Duarte, A. Sabharwal, V. Aggarwal, R. Jana, K. Ramakrishnan, C. W. Rice, and N. Shankaranarayanan, "Design and characterization of a full-duplex multiantenna system for WiFi networks," *IEEE Trans. Vehicular Technology*, vol. 63, no. 3, pp. 1160–1177, 2014.
18. E. Ahmed, A. M. Eltawil, Z. Li, and B. A. Cetiner, "Full-duplex systems using multireconfigurable antennas," *IEEE Trans. on Wireless Comm.*, vol. 14, no. 11, pp. 5971–5983, 2015.
19. A. Masmoudi and T. Le-Ngoc, "Joint radio-frequency/baseband self-interference cancellation methods and systems," 2015, US Patent App. 14/675,278.
20. ——, "Joint radio-frequency/baseband self-interference cancellation methods," 2015, US Patent App. 14/675,515.
21. ——, "A maximum-likelihood channel estimator for self-interference cancellation in full-duplex systems," *IEEE Trans. Vehicular Technology*, vol. 65, no. 7, pp. 5122–5132, July 2016.

22. T. Riihonen, S. Werner, and R. Wichman, "Mitigation of loopback self-interference in full-duplex MIMO relays," *IEEE Trans. Signal Process.*, vol. 59, no. 12, pp. 5983–5993, 2011.
23. A. Masmoudi and T. Le-Ngoc, "Self-interference mitigation using active signal injection full-duplex MIMO-OFDM systems," in *Proc. IEEE Veh. Technol. Conf.*, Montréal, Quebec, Canada, Sept. 2016.
24. E. Ahmed, A. Eltawil, and A. Sabharwal, "Simultaneous transmit and sense for cognitive radios using full-duplex: A first study," in *IEEE Antennas and Propagation Society International Symposium (APSURSI)*, Chicago, Illinois, USA.
25. S. Parsaeefard and T. Le-Ngoc, "Improving wireless secrecy rate via full-duplex relay-assisted protocols," *IEEE Trans. Information Forensics and Security*, vol. 10, no. 10, pp. 2095–2107, 2015.
26. G. Zheng, I. Krikidis, J. Li, A. P. Petropulu, and B. Ottersten, "Improving physical layer secrecy using full-duplex jamming receivers," *IEEE Trans. Signal Process*, vol. 61, no. 20, pp. 4962–4974, 2013.

Chapter 2
Self-Interference-Cancellation in Full-Duplex Systems

Abstract This chapter provides a brief overview of several important concepts related to SI-cancellation techniques to form a solid background for the following chapters. We first discuss the nature of the self-interference (SI) channel which leads to the use of the analog RF cancellation stage and the digital cancellation stage. We describe both stages and state their advantages and limitations. The next part presents a quick survey on self-interference channel estimation. We then discuss the transmitter impairments that can degrade the SI-cancellation performance and the existing methods to mitigate them. The last part presents recent advances in precoding for SI-cancellation.

2.1 SI Channel Modelling

Various measurements have been done to characterise the SI channel. Consider the simple and popular architecture using the same antenna to transmit and receive via a 3-port circulator, the dominant paths of the SI channel come from the leakage through the circulator and the internal antenna reflections due to the impedance mismatch between the isolator and the antenna. On the other hand, external reflections from closely-located objects may occur with much larger delays and weaker levels compared to the dominant paths since they travel longer distances. It was reported in [1] that the external reflections are about 30 dB lower than the leakage and antenna reflections paths. When using two different antennas to transmit and receive, the line-of-sight (LoS) components and the path coming from the electromagnetic waves reflected from the transceiver structure represent the most significant paths [2, 3]. Figure 1.4 represents the different reflections that constitute the SI channel for the two antenna configurations. In both cases, the internal reflections are static since they depend on the structure of the transceiver while the external reflections vary according to the surrounding environment. In general, the power delay profile (PDP) of the SI channel is written as [1]:

$$\mathrm{PDP}(t,\tau) = \gamma_{m_1}\delta(t-\tau_{m_1}) + \gamma_{m_2}\delta(t-\tau_{m_2}) + \sum_{l=2}^{L}\gamma_l\delta(t-\tau_l), \tag{2.1}$$

T. Le-Ngoc, A. Masmoudi, *Full-Duplex Wireless Communications Systems*, Wireless Networks, DOI 10.1007/978-3-319-57690-9_2

where $(\gamma_{m_1}, \tau_{m_1})$ and $(\gamma_{m_2}, \tau_{m_2})$ are the power/delay of the internal/coupling reflections and (γ_l, τ_l), for $l = 2, \ldots, L$ are the power/delay of the external reflections.

2.2 Analog RF Cancellation Stage

There are extensive works that describe the analog RF cancellation. Traditionally, the analog RF cancellation uses the knowledge of the transmitted SI to cancel it before the receive LNA. A copy of the transmitted signal is obtained from the PA output and passed through a cancelling circuit to reconstruct a copy of the received SI. The signal at the PA output includes the distortions of the transmitter, which are reduced by the analog RF cancellation stage.

The design of the cancelling circuit is highly related to the nature of the SI channel. As discussed in Sect. 2.1, the SI channel can be divided into internal reflections with a smaller number of paths, shorter delays and stronger amplitudes compared to the external (far-field) reflections. The internal reflections are static as they depend on the internal components and the structure of the transceiver, while the external reflections vary according to the surrounding environment. Since it is difficult to adapt the analog circuits with the variations of the external reflections, the analog RF cancellation stage reduces the static internal reflections. A cancelling circuit based on balanced transformer, such as a QHx220 chip, is used in [4, 5]. The chip takes the transmitted SI as input, changes its amplitude and its phase to match the received SI then subtracts the resulting signal from the received signal. This method achieves about 20 dB reduction in the received SI [4]. Another solution consists of using tapped delay lines (TDL) of variable delays and tunable attenuators to model the SI channel. The lines are then collected back, added up and the resulting signal is then subtracted from the received signal. Figure 2.1 shows the described TDL structure. Tuning algorithms are used to find the optimal coefficients for the attenuator, the phase shifter and the tunable delay line of each tap. The parameters of the circuit are adjusted to minimize the residual energy after cancellation [5, 6] or to minimize the error between the response of the circuit and the internal reflections response [7]. The SI reduction of the TDL varies from 30 to 45 dB [6, 7]. While the TDL can match the short delay of the internal reflections, the interaction between the delays and attenuators makes the tuning very complex. Also, analog RF cancellation is much more challenging for MIMO systems since it requires adapting different TDLs for each transmit-receive antenna pair.

As the analog RF cancellation can reduce the SI by a maximum of 45 dB, a large amount of SI is left to be reduced in the following cancellation stages. In particular, the external reflections need more adaptive cancellation methods, which can be done using digital signal processing.

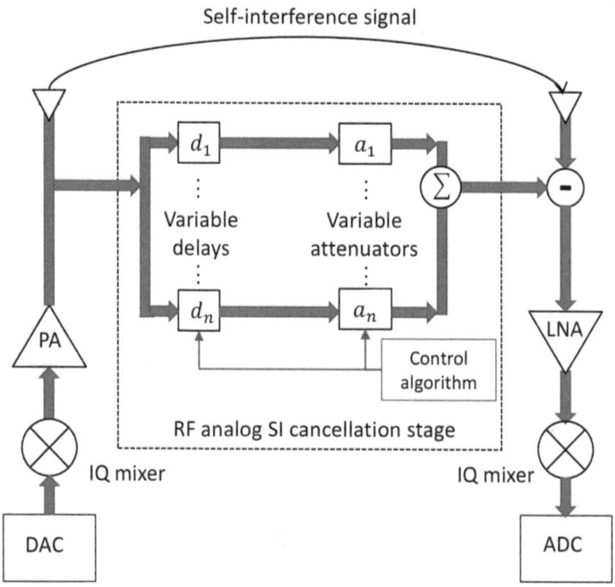

Fig. 2.1 RF analog cancellation stage

2.3 Digital SI-Cancellation

Processing the SI in the digital domain facilitates the use of adaptive digital filtering for a large number of reflected paths due to the external environment. The digital SI-cancellation is based on the general transversal symbol-synchronous finite impulse response (FIR) structure shown in Fig. 2.2, where the constant tap-delay is equal to the signal sampling period and implemented as a D-flipflop clocked by the sampling clock. Here, only the tap-coefficients need to be specified from an estimate of the SI channel and thus we avoid the interaction between the delays and the attenuations as it is the case for the analog TDL. As a result, the digital processing can deal with a larger number of taps than the analog TDL to adapt to the varying external environment. Digital SI-cancellation is particularly suitable for MIMO systems as the cross interference between antennas increases considerably the number of taps needed to reduce the SI.

The resulting cancelling signal can be subtracted from the received signal at the RF input of the LNA/ADC to further reduce the SI resulting from the external reflections and to keep the LNA/ADC not overloaded. This operation requires an additional digital-to-analog converter (DAC) and an up-converting radio chain to generate the RF signal. The additional components will slightly change the generated SI leading to residual SI. This RF cancellation stage can provide 30 dB of SI-cancellation [8, 9], which, on top of the previously-obtained 45 dB, still leaves a large amount of SI. Therefore, the baseband cancellation stage represents the last

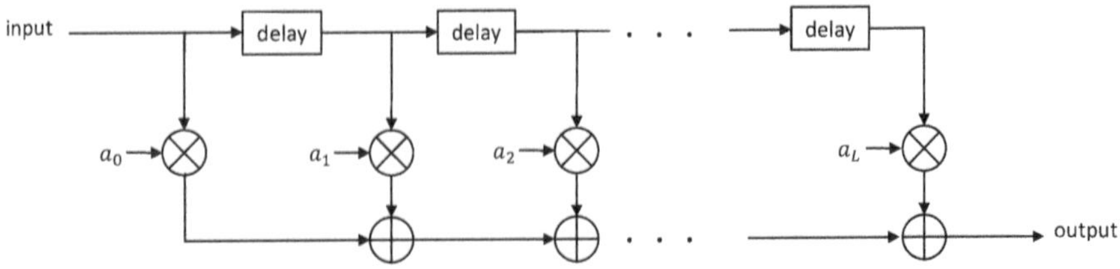

Fig. 2.2 Symbol-synchronous transversal FIR structure

line of defense against the SI by reducing it after the ADC. For this, we should estimate the transmitter nonlinearities and the residual SI channel, resulting from the difference between the actual SI channel and the equivalent channel generated by the previous cancellation stages.

2.4 Channel Estimation in Full-Duplex Systems

As previously mentioned, knowing the SI channel is an important step to reconstruct the cancelling signal. In a practical environment, it is difficult, if not impossible, to completely cancel the SI due to imperfect channel estimation [10]. In the presence of the intended signal, the received signal is expressed as:

$$y(n) = \sum_{l=0}^{L} \left(x(n-l)h^i(l) + s(n-l)h^{us}(l) \right) + w(n), \tag{2.2}$$

where $h^i(l)$ and $h^{us}(l)$ are the SI and intended channels, respectively, $(L+1)$ is the number of paths and $x(n)$ and $s(n)$ are the transmitted SI and intended signals, respectively. By collecting N observations, the resulting vector $\boldsymbol{y} = [y(1), y(2), \ldots, y(N)]^T$ is expressed as:

$$\boldsymbol{y} = \boldsymbol{X}\boldsymbol{h}^i + \boldsymbol{S}\boldsymbol{h}^{us} + \boldsymbol{w}, \tag{2.3}$$

where $\boldsymbol{X}$ and $\boldsymbol{S}$ are Toeplitz matrices obtained from the known transmitted SI signal and the unknown intended signal, respectively, $\boldsymbol{h}^i = \left[h^i(0), h^i(1), \ldots, h^i(L)\right]^T$ and $\boldsymbol{h}^{us} = [h^{us}(0), h^{us}(1), \ldots, h^{us}(L)]^T$. The existing methods follow a data-aided approach to estimate the SI channel by exploiting the knowledge of the SI data. In [8], the SI channel coefficients are obtained in the frequency domain by dividing the received signal by the known transmitted symbols over each subcarrier. A two-step Least square (LS)-based estimator is presented in [11] where a first estimate of the SI channel is obtained by considering the intended signal from the other transceiver as additive noise. After that, the interference is suppressed and the resulting signal is used to detect the intended data. A more precise estimate of the channel is then obtained by jointly estimating the SI and intended channels using the known transmitted interference and detected data. However, an initial estimate of the signal channel is important in the detection of the intended data. Minimum mean square error (MMSE) and LS channel estimators are also used in [12] and [13] for full-duplex relays and MIMO transceivers, respectively. In general, a linear estimate of $\boldsymbol{h}^i$ is given by [14]:

$$\widehat{\boldsymbol{h}}^i = \boldsymbol{M}\boldsymbol{y}, \tag{2.4}$$

where the matrix $\boldsymbol{M}$ (to be derived) determines the estimate of $\boldsymbol{h}^i$. For example, using the LS criterion, $\boldsymbol{M}$ will be given by $(\boldsymbol{X}^H\boldsymbol{X})^{-1}\boldsymbol{X}^H$, while using the MMSE criterion:

$$\boldsymbol{M} = \left(\left(\mathbb{E}\left\{\boldsymbol{h}^i\boldsymbol{h}^{iH}\right\}\right)^{-1} + \frac{1}{\sigma^2 + \sigma_s^2}\boldsymbol{X}^H\boldsymbol{X} \right)^{-1} \frac{1}{\sigma^2 + \sigma_s^2}\boldsymbol{X}^H, \tag{2.5}$$

where $\mathbb{E}\{\cdot\}$ denotes statistical expectation and σ^2 and σ_s^2 are the variances of the thermal noise and the intended signal, respectively. While the latter needs the knowledge of the second-order statistics of the SI channel, it enjoys substantially lower channel estimation error than the LS estimator.

An adaptive least mean square (LMS) algorithm to estimate the SI channel is also proposed in [15] and [16] where the large SI signal compared to the intended signal is exploited to obtain an estimate of the SI channel. However, many iterations are needed for the algorithm to converge during which it is not possible to recover the intended signal. A power allocation strategy is presented in [17] to improve the estimate of the SI channel. This strategy leads to a higher power employed to estimate the channel and less power is left for data transmission, which has the advantage of obtaining an accurate SI channel estimate but low data transmission rate.

The above-mentioned methods were motivated by the knowledge of the transmitted SI, leading to simple estimators. However, they only estimate the SI channel, making the intended signal behaves as additive noise. This increases the overall noise during the estimation process and will ultimately degrade the performance of these methods which limits the cancellation capability of full-duplex systems. Given the high requirement of the estimation accuracy, it is important to find more efficient algorithms that can estimate the desired SI channel without being affected by the intended signal. One direct solution is to set a training period during which only the transceiver itself is transmitting, and thus receiving only the SI to properly estimate the SI channel. The downside of this solution is the decrease in the throughput as the two communicating transceivers need to reserve one period each in a periodic manner to update the estimated coefficients when the SI-channel changes. Another approach is to incorporate the intended signal in the estimation process by jointly estimating the SI and intended channels. As the intended signal is unknown, the intended channel can be estimated blindly or semi-blindly if some pilot symbols are transmitted from the intended transceiver. Moreover, in most of the above-mentioned methods, the impairments of the RF components have not been considered and their effects cannot be cancelled which leaves a large amount of SI. As will be discussed in Chap. 3, reducing the transmitter impairments is primordial to properly detect the intended signal, and thus they have to be estimated and cancelled.

Spatial domain cancellation attempts to reduce the SI by precoding at the transmit chain and decoding at the receive chain [18, 19]. As detailed in Sect. 2.7, these techniques require the knowledge of both the SI and intended channels which motivates more the development of channel estimators for full-duplex systems.

2.5 Effect of Transmitter Impairments in SI-Cancellation

In order to subtract the received SI, we need to capture every modification that can occur to the transmitted SI. This includes the propagation channel and the responses of the transceiver components such as the IQ mixer and the PA. Actually, the transmitted SI is slightly modified as it goes through the transmit chain. While these modifications are relatively low compared to the main signal, they are of significant magnitude compared to the intended signal and thus will limit the performance of the full-duplex system. In most practical implementations, the inband image resulting from the transmit IQ mixer is about 30 dB lower than the direct signal [20]. In the presence of strong SI of about 50 dB higher than the intended signal, this IQ image represents additional interference for the intended signal and has to be also reduced. Several recent studies have been performed to analyze a selection of the transceiver component's impairments in the particular context of full-duplex [21–26]. We mention here that alternative high-speed DAC provides a direct conversion of the baseband signal to the RF frequency, resulting in an architecture known as a direct RF transmitter. This approach can avoid many distortions related to the up-conversion. Until now, the high-speed DACs have been only used for low frequency transmission or military communications.

In [23, 25], it was observed that the phase noise generated by the local oscillators can potentially limit the SI-cancellation capability when independent oscillators are used in the up-conversion and down-conversion. A shared-oscillator can reduce the phase noise effects and improve the cancellation performance by 25 dB [26]. In this case, the difference between the phase noise affecting the transmitted and received SI depends on the propagation delay that the SI experiences from the transmit to receive chains. A comprehensive analysis of the transceiver impairments that does not include the phase noise effects is provided in [21] and showed that nonlinear cancellation techniques should be implemented to properly reduce the SI. Such techniques can reduce the effects of the PA in the baseband cancellation stage by estimating the nonlinear coefficients of the PA [27] and another technique has been proposed to deal with the IQ mixer imbalance in [28].

2.6 Cancellation of Nonlinear Distortions

Transmitter imperfections, including the PA nonlinearity and the IQ imbalance, are significant limiting factors that bound the SI cancellation capability. To reduce these impairments, their effects should be properly modeled. The response of the PA is usually approximated by a Hammerstein model as [29, 30]:

$$x^{PA}(t) = \left(\sum_{p=0}^{P} \alpha_{2p+1} x(t)|x(t)|^{2p} \right) \star f(t), \tag{2.6}$$

where α_{2p+1}, for $p = 0, \ldots, P$, are the complex-valued polynomial coefficient for a nonlinearity order of P, and $f(t)$ is the memory of the PA. In (2.6), $\star$ denotes the convolution operator.

An iterative technique is proposed in [27] to jointly estimate the SI channel and the nonlinearity coefficients required to suppress the distorted signal. The analysis in [27] is limited to memoryless PA (i.e., $f(t) = \delta(t)$) and to the third-order nonlinearity (considers only α_3) simplifying (2.6) into:

$$x^{PA}(t) = x(t) + \alpha_3 x(t)|x(t)|^2. \tag{2.7}$$

Considering a multipath propagation channel, the received signal at the ADC output is written as:

$$y(n) = \sum_{l=0}^{L} \left(h^i(l)x(n-l) + h^{us}(l)s(n-l) \right) + d(n) + w(n), \tag{2.8}$$

where $d(n)$ collects the PA nonlinearity and the SI channel and $w(n)$ is the additive Gaussian noise. In [27], an iterative estimation technique is proposed by following these steps:

1. An initial estimate of the SI channel is obtained using the LS criterion and considering $\sum_{l=0}^{L} h^{us}(l)s(n-l) + d(n)$ as additive noise.
2. The previous estimated channel is used to find the nonlinear coefficients.
3. $d(n)$ is reconstructed and subtracted from the received signal to estimate again the SI channel.

The SI image resulting from the IQ imbalance can be attenuated by using a widely-linear representation of the received signal [22]. Actually, the output of an IQ mixer is:

$$x^{IQ}(t) = g_1 x(t) + g_2 x^*(t), \tag{2.9}$$

where g_1 and g_2 represent the response to the direct signal and the image signal, respectively. Using (2.9) to model the transmitter and the receiver IQ mixers, the discrete-time received signal is given by:

$$\begin{aligned} y(n) &= \sum_{l=0}^{L} h^i(l) x^{IQ}(n-l) + r(n) \\ &= \sum_{l=0}^{L} h_1^i(l) x(n-l) + h_2^i(l) x^*(n-l) + r(n), \end{aligned} \tag{2.10}$$

where $r(n)$ denotes the sum of all other signal, including the intended signal from the other transceiver, the thermal noise and the PA-induced nonlinearity; $h_1^i(l)$ is the equivalent SI channel of the transmitted signal $x(n)$; and $h_2^i(l)$ is the equivalent channel of the image signal $x^*(n)$ resulting from the IQ imbalance. The authors of [28] observed that the IQ imbalance can be mitigated if an estimate of $h_2^i(l)$ is available. Therefore, they estimate both $h_1^i(l)$ and $h_2^i(l)$ from the observed signal $y(n)$ based on the known transmitted signal $x(n)$ and its complex conjugate $x^*(n)$. Gathering N observations, the resulting vector $\boldsymbol{y} = [y(1), \ldots, y(N)]^T$ is expressed as:

$$\begin{aligned} \boldsymbol{y} &= \boldsymbol{X}\boldsymbol{h}_1^i + \boldsymbol{X}^*\boldsymbol{h}_2^i + \boldsymbol{r} \\ &= \underbrace{[\boldsymbol{X}\,\boldsymbol{X}^*]}_{\boldsymbol{X}_{aug}} \underbrace{\begin{pmatrix} \boldsymbol{h}_1^i \\ \boldsymbol{h}_2^i \end{pmatrix}}_{\boldsymbol{h}_{aug}} + \boldsymbol{r}, \end{aligned} \tag{2.11}$$

where $\boldsymbol{X}$ is a Toeplitz matrix obtained from the known transmitted SI signal, $\boldsymbol{X}^*$ is its complex conjugate, $\boldsymbol{h}_1^i = \left[h_1^i(0), h_1^i(1), \ldots, h_1^i(L)\right]^T$ and $\boldsymbol{h}_2^i = \left[h_2^i(0), h_2^i(1), \ldots, h_2^i(L)\right]^T$. Using these notations, the LS estimator of $\boldsymbol{h}_{aug}$ is obtained as:

$$\widehat{\boldsymbol{h}}_{aug} = (\boldsymbol{X}_{aug}^H \boldsymbol{X}_{aug})^{-1} \boldsymbol{X}_{aug}^H \boldsymbol{y}, \tag{2.12}$$

where the use of the reference signal and its complex conjugate is referred to as widely-linear estimation.

A more general approach presented in [31] takes into account the effects of both PA nonlinearities and IQ imbalance. The proposed estimator in [31] is similar to the one in [28] and the estimated vector containing the SI channel and the nonlinear parameters is given by:

$$\widehat{\boldsymbol{h}} = [\boldsymbol{X}_{\mathrm{PA}}\,\boldsymbol{X}_{\mathrm{PA}}^*]^{\#} \boldsymbol{y}, \tag{2.13}$$

where X_{PA} is a concatenation of P Toeplitz matrices with elements $x(n)|x(n)|^{2p}$, for $p = 0, \ldots, P$, where P is the polynomial order of the PA and the operator $(\cdot)^{\#}$ denotes the pseudo-inverse of a given matrix.

The aforementioned approaches to estimate the transmitter nonlinearities rely on linear estimators. Here too the intended signal has been ignored or treated as additive noise. As such, it should be expected that including the intended signal in the estimation process should provide significant performance increases. This could be related to blind channel estimation when the intended signal is unknown or semi-blind estimation when using the pilot symbols in conjunction with the unknown symbols.

Despite the extensive study on blind and semi-blind approaches in half-duplex transmission, very little effort has been made to apply them to parameter estimation in full-duplex systems. Due to the differences in the received signal structure, the blind and semi-blind algorithms developed for half-duplex transmission cannot be applied in full-duplex due to the following reasons. Actually, the presence of the SI and intended signal makes the number of parameters to estimate larger than in half-duplex. Also, we need to reduce the transmitter impairments, which imposes a different estimation strategy. Hence, there is a need to develop new estimation algorithms that are well-adapted to the full-duplex system model. This will be treated in the following chapters.

2.7 Spatial Cancellation

Spatial cancellation, based on transmit beamforming, can be combined with the previous cancellation approaches to further reduce the SI. The signal model in this section is built upon frequency-flat channels resulting from OFDM transmission over a multipath channel. We consider that the full-duplex transceiver is equipped with N_t transmit and N_r receive antennas. The received signal can be modeled as:

$$y = H^i x + H^{us} s + w, \tag{2.14}$$

where $x = [x_1, \ldots, x_{N_t}]^T$ and $s = [s_1, \ldots, s_{N_t}]^T$ are the transmitted SI and intended signal from all antennas, respectively, H^i and H^{us} are $N_r \times N_t$ matrices representing the respective MIMO channel from the transmit stream to the receive steam of the same transceiver and between the two transceivers, and w is the $N_r \times 1$ vector collecting the thermal noise. To exploit the DoF provided by the spatial domain, the transceiver applies a $N_t \times \widetilde{N}_t$ transmit precoding matrix G_{Tx} and a $\widetilde{N}_r \times N_r$ receive decoding matrix G_{Rx} with $\widetilde{N}_t \leq N_t$ and $\widetilde{N}_r \leq N_r$ being the number of input and output dimensions (or the number of independent streams), respectively. The target is to make the received SI close to zero. Thus, the transmit signal can be pre-processed using $\boldsymbol{G}_{Tx}$ as $\boldsymbol{x} = \boldsymbol{G}_{Tx}\widetilde{\boldsymbol{x}}$ and the received signal can be post-processed using $\boldsymbol{G}_{Rx}$ to obtain:

$$\begin{aligned}\widetilde{\boldsymbol{y}} &= \boldsymbol{G}_{Rx}\boldsymbol{y} \\ &= \boldsymbol{G}_{Rx}\boldsymbol{H}^{i}\boldsymbol{G}_{Tx}\widetilde{\boldsymbol{x}} + \boldsymbol{G}_{Rx}\boldsymbol{H}^{us}\boldsymbol{s} + \boldsymbol{G}_{Rx}\boldsymbol{w}.\end{aligned} \tag{2.15}$$

Roughly speaking, the processing matrices $\boldsymbol{G}_{Tx}$ and $\boldsymbol{G}_{Rx}$ modify the true SI channel seen by the receiver. Here, these two matrices are designed to reduce the SI given by $\boldsymbol{G}_{Rx}\boldsymbol{H}^{i}\boldsymbol{G}_{Tx}\widetilde{\boldsymbol{x}}$. One technique, called antenna selection[1] (AS) [19], selects the transmit and receive antenna pairs which lead to a minimal received SI. To that end, the transmit and receive filters are implemented to minimize[2]:

$$||\boldsymbol{G}_{Rx}\boldsymbol{H}^{i}\boldsymbol{G}_{Tx}||_F^2, \tag{2.16}$$

with[3] the additional constraint that $\boldsymbol{G}_{Rx}^T$ and $\boldsymbol{G}_{Tx}$ are two subset selection matrices (i.e., matrices with binary elements such that $\sum_i \boldsymbol{G}_{Tx}(i,j) = 1$ for all j and $\sum_j \boldsymbol{G}_{Tx}(i,j) \in \{0, 1\}$ for all i). The filters that minimize (2.16) are obtained by calculating the Frobenius norm for all possible combinations and choosing the lowest.

The beam selection (BS) technique is based on the singular value decomposition (SVD) of the matrix $\boldsymbol{H}^{i}$ to find the transmit and receive filters [19, 32]. More specifically, by writing $\boldsymbol{H}^{i} = \boldsymbol{U}\boldsymbol{\Sigma}\boldsymbol{V}^H$, where $\boldsymbol{U}$ and $\boldsymbol{V}$ are unitary matrices and the diagonal matrix $\boldsymbol{\Sigma}$ comprises the singular values of $\boldsymbol{H}^{i}$, the BS is performed by first finding the subset selection matrices $\boldsymbol{S}_{Rx}$ and $\boldsymbol{S}_{Tx}$ that minimize:

$$||\boldsymbol{S}_{Rx}^T\boldsymbol{\Sigma}\boldsymbol{S}_{Tx}||_F^2. \tag{2.17}$$

Then, the BS matrices are chosen as:

$$\boldsymbol{G}_{Tx} = \boldsymbol{V}\boldsymbol{S}_{Tx} \text{ and } \boldsymbol{G}_{Rx} = \boldsymbol{S}_{Rx}^T\boldsymbol{U}^H. \tag{2.18}$$

The row and column selection in the BS is based on the diagonal matrix $\boldsymbol{\Sigma}$ such that the subset selection matrices $\boldsymbol{S}_{Rx}$ and $\boldsymbol{S}_{Tx}$ are chosen to select the off-diagonal elements of $\boldsymbol{\Sigma}$.

The null-space projection (NSP) has been proposed to completely eliminate the SI [19, 33–35]. In this method, $\boldsymbol{G}_{Tx}$ and $\boldsymbol{G}_{Rx}$ are selected such that the transmission and reception are performed in different subspaces, i.e., the transmit signal is

[1]This method was originally proposed for full-duplex relay station.

[2]In practical implementation, only an estimate of the SI channel is available and used in the minimization process.

[3]In (2.16), $||.||_F$ returns the Frobenius norm of a matrix.

projected into the null-space of the SI channel. Therefore, the filter design is stated as:

$$\boldsymbol{G}_{Rx}\boldsymbol{H}^{i}\boldsymbol{G}_{Tx} = \boldsymbol{0}. \tag{2.19}$$

The way to solve the NSP depends on the rank of the SI channel. If $\min\{N_t, N_r\}$ is larger than the rank of $\boldsymbol{H}^i$, the BS previously discussed provides $\boldsymbol{S}_{Rx}^T \boldsymbol{\Sigma} \boldsymbol{S}_{Tx} = \boldsymbol{0}$ by selecting the singular value *zero* when choosing $\boldsymbol{S}_{Rx}$ and $\boldsymbol{S}_{Tx}$. For general low rank SI channel, $\boldsymbol{G}_{Tx}$ can be chosen to belong to the right null space of $\boldsymbol{H}^i$ (by taking the columns of $\boldsymbol{V}$ associated with the singular value zero of $\boldsymbol{H}^i$) or similarly we choose $\boldsymbol{G}_{Rx}$ to belong to the left null space of $\boldsymbol{H}^i$. Other designs can also be adopted in the particular case of $N_t = N_r = 2$ and $\widetilde{N}_t = \widetilde{N}_r = 1$ [35].

In general, spatial cancellation requires the number of transmitting antennas to be larger than the number of receive antennas and reduces the available data stream for SI-cancellation.

2.8 Chapter Summary

This chapter provided an overview of the existing works on SI-cancellation in full-duplex wireless systems. The cancellation techniques can go from subtraction of the received SI into precoding to reduce the coupling signal. It was seen that the estimation of the SI channel is a central issue to develop efficient cancellation methods. This review provided some motivation for the material proposed in this book. In particular, we focus on the digital SI-cancellation by developing efficient estimators while the design of the RF analog cancellation is beyond the scope of the book. A CS-based estimator is presented for the RF cancellation stage, which exploits the sparsity of the SI channel. Then, subspace-based and ML algorithms are developed to estimate the residual SI and the transmitter nonlinearities for the baseband cancellation stage, in the presence of the unknown intended signal. Also, a new ASI method is proposed to substitute or complete the RF cancellation stage. Before that, a detailed study of the received SI is presented in the next chapter to help understanding its nature and developing the appropriate cancellation techniques.

References

1. X. Wu, Y. Shen, and Y. Tang, "The power delay profile of the single-antenna full-duplex self-interference channel in indoor environments at 2.6 GHz," *IEEE Antennas and Wireless Propagation Letters*, vol. 13, pp. 1561–1564, 2014.
2. C. R. Anderson, S. Krishnamoorthy, C. G. Ranson, T. J. Lemon, W. G. Newhall, T. Kummetz, and J. H. Reed, "Antenna isolation, wideband multipath propagation measurements, and interference mitigation for on-frequency repeaters," in *Proc. IEEE SoutheastCon*, Greensboro, North Carolina, USA, Mar. 2004.

3. A. Sethi, V. Tapio, and M. Juntti, "Self-interference channel for full duplex transceivers," in *Proc. IEEE Wireless Commun. and Netw. Conf.*, Istanbul, Turkey, Apr. 2014.
4. J. I. Choi, M. Jain, K. Srinivasan, P. Levis, and S. Katti, "Achieving single channel, full duplex wireless communication," in *Proc. ACM MobiCom*, Chicago, Illinois, USA, Sept. 2010.
5. M. Jain, J. I. Choi, T. Kim, D. Bharadia, S. Seth, K. Srinivasan, P. Levis, S. Katti, and P. Sinha, "Practical, real-time, full duplex wireless," in *Proc. of ACM Mobicom*, Las Vegas, Nevada, USA, Sept. 2011.
6. D. Bharadia, E. McMilin, and S. Katti, "Full duplex radios," in *Proc. of the ACM SIGCOMM 2013*, Hong Kong, China, Aug. 2013.
7. K. E. Kolodziej, J. G. McMichael, and B. Perry, "Adaptive RF canceller for transmit-receive isolation improvement," in *IEEE Radio and Wireless Symposium (RWS)*, Newport Beach, California, USA, Jan. 2014.
8. M. Duarte, C. Dick, and A. Sabharwal, "Experiment-driven characterization of full-duplex wireless systems," *IEEE Trans. Wireless Comm.*, vol. 11, no. 12, pp. 4296–4307, 2012.
9. M. Duarte, A. Sabharwal, V. Aggarwal, R. Jana, K. Ramakrishnan, C. W. Rice, and N. Shankaranarayanan, "Design and characterization of a full-duplex multiantenna system for WiFi networks," *IEEE Trans. Vehicular Technology*, vol. 63, no. 3, pp. 1160–1177, 2014.
10. D. Kim, H. Ju, S. Park, and D. Hong, "Effects of channel estimation error on full-duplex two-way networks," *IEEE Trans. Vehicular Technology*, vol. 62, no. 9, p. 4667, 2013.
11. S. Li and R. D. Murch, "Full-duplex wireless communication using transmitter output based echo cancellation," in *Proc. IEEE Global Telecommun. Conf.*, Houston, Texas, USA, Dec. 2011.
12. J. Ma, G. Y. Li, J. Zhang, T. Kuze, and H. Iura, "A new coupling channel estimator for cross-talk cancellation at wireless relay stations," in *Proc. IEEE Global Telecommun. Conf.*, Honolulu, Hawaii, USA, Nov. 2009.
13. B. P. Day, A. R. Margetts, D. W. Bliss, and P. Schniter, "Full-duplex bidirectional MIMO: Achievable rates under limited dynamic range," *IEEE Trans. Signal Process*, vol. 60, no. 7, pp. 3702–3713, 2012.
14. S. M. Kay, *Fundamentals of statistical signal processing, Volume 1: Estimation theory*. Prentice Hall, 1993.
15. J. R. Krier and I. F. Akyildiz, "Active self-interference cancellation of passband signals using gradient descent," in *Proc. IEEE Pers. Indoor and Mobile Radio Commun.*, London, UK, Sept. 2013.
16. N. Li, W. Zhu, and H. Han, "Digital interference cancellation in single channel, full duplex wireless communication," in *Prog. Wireless Communications, Networking and Mobile Computing (WiCOM)*, Shanghai, China, Sept. 2012.
17. Z. Zhan, G. Villemaud, J.-M. Gorce *et al.*, "Analysis and reduction of the impact of thermal noise on the full-duplex ofdm radio," in *IEEE Radio Wireless Symposium*, Newport Beach, California, USA, Jan. 2014.
18. D. Bliss, P. Parker, and A. Margetts, "Simultaneous transmission and reception for improved wireless network performance," in *IEEE/SP Workshop on Statistical Signal Processing*, Madison, Wisconsin, USA, Aug. 2007.
19. T. Riihonen, S. Werner, and R. Wichman, "Mitigation of loopback self-interference in full-duplex MIMO relays," *IEEE Trans. Signal Process.*, vol. 59, no. 12, pp. 5983–5993, 2011.
20. "LTE; evolved universal terrestrial radio access (E-UTRA); user equipment (UE) radio transmission and reception (3GPP TS 36.101 version 11.2.0 release 11)." ETSI, Sophia Antipolis Cedex, France.
21. D. Korpi, T. Riihonen, V. Syrjala, L. Anttila, M. Valkama, and R. Wichman, "Full-duplex transceiver system calculations: Analysis of ADC and linearity challenges," *IEEE Trans. Wireless Comm.*, vol. 13, no. 7, pp. 3821–3836, 2014.
22. A. Masmoudi and T. Le-Ngoc, "Self-interference cancellation limits in full-duplex communication systems," in *Proc. IEEE Global Telecommun. Conf.*, Washington DC, USA, Dec. 2016.

23. A. Sahai, G. Patel, C. Dick, and A. Sabharwal, "On the impact of phase noise on active cancelation in wireless full-duplex," *IEEE Trans. on Vehicular Technology*, vol. 62, no. 9, pp. 4494–4510, Nov 2013.
24. A. Masmoudi and T. Le-Ngoc, "Residual self-interference after cancellation in full-duplex systems," in *Proc. IEEE Int. Conf. Commun.*, Sydney, Australia, June 2014.
25. E. Ahmed and A. M. Eltawil, "On phase noise suppression in full-duplex systems," *IEEE Trans. on Wireless Comm.*, vol. 14, no. 3, pp. 1237–1251, 2015.
26. V. Syrjala, M. Valkama, L. Anttila, T. Riihonen, and D. Korpi, "Analysis of oscillator phase-noise effects on self-interference cancellation in full-duplex OFDM radio transceivers," *IEEE Trans. on Wireless Comm.*, vol. 13, no. 6, pp. 2977–2990, June 2014.
27. E. Ahmed, A. M. Eltawil, and A. Sabharwal, "Self-interference cancellation with nonlinear distortion suppression for full-duplex systems," in *Proc. ASILOMAR Signals, Syst., Comput.*, Pacific Frove, California, USA, Nov. 2013.
28. D. Korpi, L. Anttila, V. Syrjala, and M. Valkama, "Widely linear digital self-interference cancellation in direct-conversion full-duplex transceiver," *IEEE Journal on Selected Areas in Communications*, vol. 32, no. 9, pp. 1674–1687, Sept 2014.
29. A. Masmoudi and T. Le-Ngoc, "Channel estimation and self-interference cancellation in full-duplex communication systems," *IEEE Trans. Vehicular Technology*, vol. 66, no. 1, pp. 321–334, Jan. 2017.
30. ——, "A digital subspace-based self-interference cancellation in full-duplex MIMO transceivers," in *Proc. IEEE Int. Conf. Commun.*, London, UK, June 2015.
31. L. Anttila, D. Korpi, E. Antonio-Rodríguez, R. Wichman, and M. Valkama, "Modeling and efficient cancellation of nonlinear self-interference in MIMO full-duplex transceivers," in *IEEE Globecom Workshops*, Austin, Texas, USA, Dec. 2014.
32. P. Larsson and M. Prytz, "MIMO on-frequency repeater with self-interference cancellation and mitigation." in *Proc. of IEEE VTC-Spring*, Yokohama, Japan, May 2009.
33. B. Chun and Y. H. Lee, "A spatial self-interference nullification method for full duplex amplify-and-forward MIMO relays," in *Proc. IEEE Wireless Commun. and Netw. Conf.*, Sydney, Australia, Apr. 2010.
34. B. Chun and H. Park, "A spatial-domain joint-nulling method of self-interference in full-duplex relays," *IEEE Comm. Lett.*, vol. 16, no. 4, pp. 436–438, 2012.
35. H. Ju, E. Oh, and D. Hong, "Improving efficiency of resource usage in two-hop full duplex relay systems based on resource sharing and interference cancellation," *IEEE Trans. Wireless Comm.*, vol. 8, no. 8, pp. 3933–3938, 2009.

Chapter 3
Limiting Factors in Self-Interference Cancellation

Abstract Various experimental results have stated that the self-interference (SI) in full-duplex communication can be mitigated to properly detect the intended signal. However, relatively little is known about the cancellation limits of these systems. A good SI-cancellation strategy passes first by a proper understanding of the nature of this SI. This is highly related to the physical implementations of radio-frequency (RF) transceivers that suffer from a variety of non-idealities defying the perfect knowledge of the received SI. In this chapter, we investigate the various factors that can limit the SI-cancellation performance in full-duplex wireless communications systems. We first classify the known SI-cancellation architectures based on where the reference signal is taken to cancel the SI. By combining the effects of the transceiver impairments, estimation error and SI channel, our analysis reveals that the main bottleneck SI-cancellation turns out to be either the quantization noise, the local oscillator phase noise or the estimation error, depending on the selected SI-cancellation architecture. We provide illustrative numerical results that can facilitate the SI-cancellation architecture selection to meet the specific target system requirements.

In the previous chapters, we briefly discussed the need to reduce the SI in full-duplex systems. We also explained that due to the high power of the SI, successive cancellation stages are needed to properly detect the intended signal. In general, it is difficult to give the exact amount of the SI reduction that can be obtained because of the interactions between many factors such as the transceiver impairments, the wireless propagation channel, the estimation error, etc. What can be done is the identification of the main factors that affect the cancellation performance. This allows a better understanding of the obtained performance and eventually to develop new methods to improve the cancellation capability in full-duplex systems.

In this chapter, we address in detail the impact of the transceiver impairments in full-duplex systems. The analysis characterizes the residual SI after the RF and baseband cancellation stages and specifies the limiting factors of each stage. To that end, explicit expressions are provided to quantify the distortion power caused by the RF components at different points in the receiver.

T. Le-Ngoc, A. Masmoudi, *Full-Duplex Wireless Communications Systems*,
Wireless Networks, DOI 10.1007/978-3-319-57690-9_3

Depending on where the reference signal is taken from, the RF and baseband cancellation stages may or may not reduce the transceiver impairments. This leads to different architectures of the full-duplex transceiver. The reference signal used for the RF cancellation stage can be (1) taken from the baseband and then up-converted to RF by an auxiliary transmitter chain, or (2) taken directly from the output of the PA through a coupler. For the baseband cancellation stage, (1) an auxiliary receiver chain can be used to obtain a baseband reference signal from the output of the PA and used to cancel the SI or (2) the reference signal can be directly taken from the output of the modulator. As it will be discussed, each architecture has its own limitations.

The effect of transceiver impairments has also been analyzed in [1, 2] but without covering the different possible architectures and by ignoring the phase noise. Also, the baseband cancellation stage in [1] reduces only the linear part of the SI, which does not reflect the actual performance that can be obtained when applying nonlinear cancellation. In the following, we introduce the phase noise when analysing the received SI to obtain a more realistic picture about the residual SI. We focus on the scenario where the transmit and receive chains have a common oscillator which reduces the phase noise compared to using separate-oscillators [3]. Moreover, a common oscillator is a more natural choice for compact full-duplex transceiver since the transmission and reception are performed on the same frequency.

The organization of this chapter is as follows. Section 3.1 describes the full-duplex transceiver model with an emphasis on the impairment characteristics. For simplicity and clarity in presentation, the discussions focus on a single-input single-output (SISO) transceiver. However, extension to MIMO transceiver is straightforward. Section 3.2 analyses the powers of the intended signal and the interfering signal components in different stages of the receiver. Conclusions are drawn in Sect. 3.3.

3.1 System Model and SI-Cancellation Architectures

We consider two transceivers communicating in a full-duplex fashion by simultaneously transmitting and receiving over the same frequency slot. In addition to the intended signal, each transceiver receives its own SI that needs to be suppressed/cancelled for proper detection of the intended signal. Figure 3.1 shows the FD transceiver under consideration, which follows a typical direct-conversion architecture [4–6]. Our main focus is to identify the limiting factors on the SI-cancellation performance. In general, the SI can be estimated to create an SI replica (by the RF SI-conciliation stage) to be subtracted from the RF received signal at the input of the Rx LNA. Residual SI can be further estimated to create a residual SI replica (by the baseband SI-cancellation stage) to be subtracted from the baseband received signal prior to the demodulation. Since the intended received signal suffers a large propagation loss, it is much smaller than the SI (e.g., by 100 dB depending on the distance), the RF SI-cancellation stage is required to suppress the inband SI to

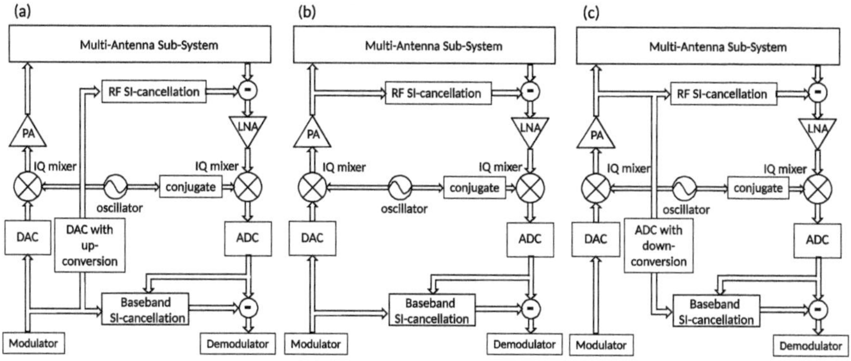

Fig. 3.1 Full-duplex transceiver block-diagrams with RF and baseband SI-cancellation stages. (**a**) First architecture. (**b**) Second architecture. (**c**) Third architecture

a sufficiently low level in order to prevent the LNA from saturation. The baseband SI-cancellation stage may be needed to further suppress the inband residual SI to the thermal noise level in order to avoid detection performance degradation. One significant aspect is the reference signal for the RF and baseband cancellation stages. In the following, we analyze three widely-used architectures. In the first architecture, the reference signals for both the RF and baseband stages are taken from the baseband transmit signal. The SI replica is first created at baseband using digital processing, and then converted to RF (for RF cancellation stage) by using an auxiliary Tx chain (including an additional set of DAC, up-conversion and PA). In the first architecture, while allowing a simple baseband digital signal, the reference signal taken from the transmit baseband signal does not include Tx impairments due to Tx PA nonlinearity and noise embedded in the SI in the creation of the SI replica. In the second architecture, the reference signal for the RF SI-cancellation stage is taken from the output of the PA and the baseband stage reference signal is taken from the baseband transmit signal. The third architecture takes the reference signals for both the RF and baseband cancellation stages from the PA output. Therefore, an auxiliary Rx chain is needed to obtain the baseband reference signal. As the reference signal taken from the PA output (in the second and third architectures) also contains the different Tx-generated impairments (nonlinearity and noise), it is possible to reduce both the SI and embedded Tx impairments. This copy of the signal is then passed through a circuit which consists of parallel lines of variable attenuators and delays. These lines are implemented to mimic the SI channel in order to obtain a copy of the received SI. The resulting signal is then subtracted from the received signal. For a multipath SI channel, we need a correspondingly-large number of delay lines, which becomes quickly limited due to various reasons such as space limitations and power consumption. In this case, only the main paths are reduced. On the other hand, by generating the RF cancellation signal from the digital transmitted samples, the cancelling signal is processed in the digital domain, which decreases the complexity of the RF circuitry.

The received signal at the antenna input can be written as:

$$y(t) = \sum_{l=0}^{L-1} h_l^{\text{si}} z(t-\tau_l) e^{j(w_c t+\phi(t-\tau_l))} + x(t)e^{jw_c t} + n_{\text{Tx}}(t) + n(t), \tag{3.1}$$

where $z(t)$ is the transmitted SI containing the known linear part $z_{\text{linear}}(t)$ and the Tx impairments $z_{\text{imp}}(t)$ $\big($i.e., $z(t) = z_{\text{linear}}(t) + z_{\text{imp}}(t)\big)$, $x(t)$ is the received signal from the other intended transceiver, h_l^{si} and τ_l denote, respectively, the attenuation and delay of the lth multipath component of the SI channel, with L being the number of paths of the SI channel and $n(t)$ and $n_{\text{Tx}}(t)$ are the thermal noise and the Tx-generated noise, respectively. In (3.1), $w_c = 2\pi f_c$ is the angular carrier frequency and $\phi(t)$ is the phase noise affecting the transmitted signal at time t. Notice that the phase noise affecting the intended signal is not considered, since the focus of this work is on the SI signal.

Considering the first architecture, the cancelling signal is obtained by convolving the transmitted baseband samples $z(n)$ with an estimate of the SI channel $\widehat{h}^{\text{si}}(n)$ then up-converting the result to obtain the equivalent signal in the RF as:

$$\widehat{y}_{\text{si}}(t) = \sum_{l=0}^{L-1} \widehat{h}_l^{\text{si}} z_{\text{linear}}(t-\widehat{\tau}_l) e^{j(w_c t+\phi(t))} + n_{\text{aux}}(t), \tag{3.2}$$

where $\widehat{h}_l^{\text{si}}$ and $\widehat{\tau}_l$ are estimates of the attenuation and the delay of the lth path and $n_{\text{aux}}(t)$ is the additive noise generated by the auxiliary Tx chain. In (3.2), the phase noise process in the auxiliary chain is the same as the phase noise on the main Tx chain as the same common oscillator is used. We mention that, since the transmitted SI in (3.1), multiplied by a time-varying phase noise process, are further convolved by the multipath SI channel, the received signal at time t is affected by relayed versions of the phase noise. While the cancelling signal in (3.2) is affected by the phase noise at time t. Thus, after subtracting the cancelling signal at the RF stage, the received signal $y_{\text{RF}}(t) = y(t) - \widehat{y}_{\text{si}}(t)$ is expressed as:

$$\begin{aligned} y_{\text{RF}}(t) &= \sum_{l=0}^{L-1} h_l^{\text{si}} z_{\text{linear}}(t-\tau_l) e^{j(w_c t+\phi(t-\tau_l))} - \widehat{h}_l^{\text{si}} z_{\text{linear}}(t-\widehat{\tau}_l) e^{j(w_c t+\phi(t))} \\ &+ h_l^{\text{si}} z_{\text{imp}}(t-\tau_l) e^{j(w_c t+\phi(t-\tau_l))} + x(t)e^{jw_c t} + n_{\text{Tx}}(t) + n_{\text{aux}}(t) + n(t). \end{aligned} \tag{3.3}$$

On the other hand, the RF cancellation stage is implemented in the same way for both the second and third architectures where the cancelling signal is given by:

$$\widehat{y}_{\text{si}}(t) = \sum_{l=0}^{L_{\text{RF}}-1} \widehat{h}_l^{\text{si}} z(t-\widehat{\tau}_l) e^{j(w_c t+\phi(t-\widehat{\tau}_l))}, \tag{3.4}$$

where L_{RF} paths are generated in the RF cancellation stage.

3.1.1 Complex-Baseband Signal Representation

After down-conversion, the complex-baseband representation of the received signal, in the first architecture, is obtained by multiplying the RF signal $y_{\mathrm{RF}}(t)$ in (3.3) by $e^{-j(w_c t+\phi(t))}$ resulting in:

$$y_{\mathrm{BB}}(t) = \sqrt{g_{\mathrm{Rx}}}\left(\sum_{l=0}^{L-1} h_l^{\mathrm{si}} z_{\mathrm{linear}}(t-\tau_l)e^{j(\phi(t-\tau_l)-\phi(t))} - \widehat{h}_l^{\mathrm{si}} z_{\mathrm{linear}}(t-\widehat{\tau}_l)\right.$$
$$\left. + h_l^{\mathrm{si}} z_{\mathrm{imp}}(t-\tau_l)e^{j(\phi(t-\tau_l)-\phi(t))} + x(t) + n_{\mathrm{aux}}(t)\right) + n_{\mathrm{imp,Rx}}(t) + n(t), \quad (3.5)$$

where g_{Rx} is the linear gain of the Rx chain and $n_{\mathrm{imp,Rx}}(t)$ collects the different nonlinearities introduced by the receiver, mainly from the receive LNA and IQ mixer. To highlight the contribution of the phase noise on the received SI, we suppose a small variation of the phase noise during the propagation delays such that the difference $\phi(t-\tau_l)-\phi(t)$ is small.[1] Thus, considering the approximation $e^{j(\phi(t-\tau_l)-\phi(t))} \approx 1 + j\left[\phi(t-\tau_l)-\phi(t)\right]$ and using the notation $\phi(t-\tau_l)-\phi(t) \triangleq \delta_\phi(t,\tau_l)$, the received signal in (3.5) can be approximated by:

$$y_{\mathrm{BB}}(t) \approx \sqrt{g_{\mathrm{Rx}}}\left(\sum_{l=0}^{L-1} h_l^{\mathrm{si}} z_{\mathrm{linear}}(t-\tau_l) - \widehat{h}_l^{\mathrm{si}} z_{\mathrm{linear}}(t-\widehat{\tau}_l) + j\delta_\phi(t,\tau_l) h_l^{\mathrm{si}} z_{\mathrm{linear}}(t-\tau_l)\right.$$
$$\left. + \left[1 + j\delta_\phi(t,\tau_l)\right] h_l^{\mathrm{si}} z_{\mathrm{imp}}(t-\tau_l) + x(t) + n_{\mathrm{aux}}\right) + n_{\mathrm{imp,Rx}}(t) + n(t). \quad (3.6)$$

This model captures the essential effects of the SI channel error after the RF cancellation stage, the transmitter and Rx nonlinearities and the effect of the phase noise on the received SI given by the term[2] $\sum_{l=0}^{L-1} j\delta_\phi(t,\tau_l) h_l^{\mathrm{si}}\left(z_{\mathrm{linear}}(t-\tau_l) + z_{\mathrm{imp}}(t-\tau_l)\right)$. In the following analysis, all the distortions and noises are modeled as additive terms [4, 7]. The power of the residual SI after the RF cancellation stage for the first architecture is expressed as:

$$p_{\mathrm{si,RF}} = \left(\frac{p_{\mathrm{si,in}}}{\beta_{\mathrm{RF}}} + p_{\mathrm{nl,Tx}} + p_{\mathrm{pn}} + p_{\mathrm{nl,aux,Rx}} + p_{\mathrm{noise,Tx}}\right) g_{\mathrm{Rx}} + p_{\mathrm{nl,Rx}}, \quad (3.7)$$

[1]The phase noise can be modeled as a Wiener process where the difference $\phi(t-\tau_l)-\phi(t)$ follows a Gaussian process with zero mean and variance $4\pi f_{3\mathrm{dB}}\tau_l$. Therefore, the probability that $|\phi(t-\tau_l)-\phi(t)| > 0.05$ is equal to 10^{-12} for practical oscillators with $f_{3\mathrm{dB}} = 10\,\mathrm{Hz}$ and $\tau_l = 400\,\mathrm{ns}$.

[2]Strictly speaking, $\sum_{l=0}^{L-1} j\delta_\phi(t,\tau_l) h_l^{\mathrm{si}} z_{\mathrm{imp}}(t-\tau_l)$ results from combination effects of Tx nonlinearities and phase noise.

where $p_{\text{si,in}}$ is the SI power of the linear part at the input of the Rx after the antenna cancellation stage given by $\sum_{l=0}^{L-1} h_l^{\text{si}} z_{\text{linear}}(t-\tau_l)$, and $p_{\text{nl,Tx}}$, $p_{\text{nl,Rx}}$ and $p_{\text{nl,aux,Rx}}$ are the power of the nonlinear distortions produced by the Tx chain, the Rx chain and the additional auxiliary chain used for the RF cancellation, respectively. In (3.7), β_{RF} represents the amount of suppression achieved by the RF cancellation stage, which, as will be shown later, depends on the estimated SI channel and the local oscillator quality (or, equivalently, the phase noise variance) and p_{pn} is the power of the phase noise-induced SI. In (3.7), β_{RF} is applied to the linear part of the SI. We choose to express the individual contribution of every component in the residual SI in order to identify the factors that limit the cancellation performance. The power of the residual SI at the input of the ADC will also be used to derive the quantization noise[3] introduced by the ADC. Actually, the ADC is preceded by a VGA, with adjustable gain g_{VGA}, so that the input signal fits the operating range of the ADC. As a result, the presence of strong interference in the received signal effectively decreases the number of bits usable for the signal of interest. In our development, we assume that the VGA operates in an optimal manner to minimize the clipping probability and the quantization noise.

After ADC and sampling the baseband signal at instant $t = nT_s$, with T_s being the sampling period, we can express (3.6) as:

$$
\begin{aligned}
y_{\text{BB}}(n) \approx \sqrt{g_{\text{Rx}}}\Bigg(& \sum_{l=0}^{L-1} h^{\text{si}}(l) z_{\text{linear}}(n-l) - \widehat{h}^{\text{si}}(l) z_{\text{linear}}(n-l) \\
& + h^{\text{si}}(l) z_{\text{imp}}(n-l) + j\delta_{\phi}(n,l) h^{\text{si}}(l) \big(z_{\text{linear}}(n-l) + z_{\text{imp}}(n-l)\big) \\
& + x(n) + n_{\text{aux}}(n) \Bigg) + n_{\text{imp,Rx}}(n) + n_{\text{ADC}}(n) + n(n), \qquad (3.8)
\end{aligned}
$$

where $n_{\text{ADC}}(n)$ denotes the quantization noise introduced by the ADC.

Following the same procedure, the received signal for the second and third architectures, after down-conversion, can be written as:

$$
\begin{aligned}
y_{\text{BB}}(t) = \sqrt{g_{\text{Rx}}}\Bigg(& \sum_{l=0}^{L_{\text{RF}}-1} h_l^{\text{si}} z(t-\tau_l) e^{j(\phi(t-\tau_l)-\phi(t))} - \widehat{h}_l^{\text{si}} z(t-\widehat{\tau}_l) e^{j(\phi(t-\widehat{\tau}_l)-\phi(t))} \\
& + \sum_{l=L_{\text{RF}}}^{L-1} h_l^{\text{si}} z(t-\tau_l) e^{j(\phi(t-\tau_l)-\phi(t))} + x(t) \Bigg) + n_{\text{imp,Rx}}(t) + n(t) \\
\overset{(a)}{\approx} \sqrt{g_{\text{Rx}}}\Bigg(& \sum_{l=0}^{L_{\text{RF}}-1} h_l^{\text{si}} z(t-\tau_l) - \widehat{h}_l^{\text{si}} z(t-\widehat{\tau}_l) + j\delta_{\phi}(t,\tau_l) h_l^{\text{si}} z(t-\tau_l)
\end{aligned}
$$

[3]The quantization noise includes also the clipping noise from the lowest and highest signal values at the output of the ADC.

$$-j\delta_\phi(t,\widehat{\tau}_l)\widehat{h}_l^{\text{si}}z(t-\widehat{\tau}_l) + \sum_{l=L_{\text{RF}}}^{L-1} h_l^{\text{si}}z(t-\tau_l)[1+j\delta_\phi(t,\tau_l)] + x(t)\Bigg)$$

$$+ n_{\text{imp,Rx}}(t) + n(t), \quad (3.9)$$

where $\overset{(a)}{\approx}$ follows from the approximation $e^{j(\phi(t-\tau_l)-\phi(t))} \approx 1 + j[\phi(t-\tau_l) - \phi(t)]$. From (3.9), the residual SI resulting from the phase noise is given by:

$$\sum_{l=0}^{L_{\text{RF}}-1} j\delta_\phi(t,\tau_l)h_l^{\text{si}}z(t-\tau_l) - j\delta_\phi(t,\widehat{\tau}_l)\widehat{h}_l^{\text{si}}z(t-\widehat{\tau}_l) + \sum_{l=L_{\text{RF}}}^{L-1} j\delta_\phi(t,\tau_l)h_l^{\text{si}}z(t-\tau_l), \quad (3.10)$$

where the phase noise affecting the L_{RF} cancelled paths in the RF cancellation stage can be completely mitigated if the corresponding SI channel coefficients are perfectly estimated. On the other hand, the phase noise appears as a limiting factor for the first architecture as a perfect estimate of the SI channel cannot reduce the effect of phase noise.

Taking into account the received signal in (3.9), the power of the residual SI after the RF stage, for both second and third architectures, can be expressed as:

$$p_{\text{si,RF}} = \left(\frac{p_{\text{si,in}}^{(1)} + p_{\text{nl,Tx}} + p_{\text{noise,Tx}}}{\beta_{\text{RF}}} + p_{\text{si,in}}^{(2)} + p_{\text{pn}}\right) g_{\text{Rx}} + p_{\text{nl,Rx}}, \quad (3.11)$$

where the power of the received SI is divided in two parts:

- $p_{\text{si,in}}^{(1)}$ represents the SI whose corresponding paths, given by $\sum_{l=0}^{L_{\text{RF}}-1} h_l^{\text{si}} z_{\text{linear}}(t-\tau_l)$, are generated and cancelled in the RF cancellation stage.
- $p_{\text{si,in}}^{(2)}$ represents the received SI from the other paths $\sum_{l=L_{\text{RF}}}^{L-1} h_l^{\text{si}} z(t-\tau_l)$ that are not reduced in the RF cancellation stage.

The performance of the RF stage, in this case, depends on the number of delays and attenuators that are physically implemented. These architectures reduces the SI coming from the most significant paths and leaves the weaker paths to be reduced in the baseband.

The received signal in the baseband, for the second and third architectures, is written as:

$$y_{\text{BB}}(n) = \sqrt{g_{\text{Rx}}}\Bigg(\sum_{l=0}^{L_{\text{RF}}-1}\left(h^{\text{si}}(l)z(n-l) - \widehat{h}^{\text{si}}(l)z(n-l)\right)[1+j\delta_\phi(n,l)]$$

$$+ \sum_{l=L_{\text{RF}}}^{L-1} h^{\text{si}}(l)z(n-l)\left[1+j\delta_\phi(n,l)\right] + x(n)\Bigg)$$

$$+ n_{\text{imp,Rx}}(n) + n_{\text{ADC}}(n) + n(n). \quad (3.12)$$

It is noted that the effects of the phase noise on the intended signal, as well as the intended channel estimation, are not discussed in this work as the objective is to study the SI.

3.1.2 *Residual SI Due to Phase Noise*

The presence of the phase noise leads to a residual SI term that cannot be reduced by linear or nonlinear cancellation techniques. This residual SI may change depending on the transceiver architecture. For the first architecture, the power of the phase noise-induced residual SI is given by:

$$p_{\text{pn}} = p_{\text{si,Tx}} \sum_{l=0}^{L-1} \alpha_l \sigma^2_{\text{pn},l}, \tag{3.13}$$

where $\sigma^2_{\text{pn},l} = 4\pi f_{3\,\text{dB}} T_s l$ is the variance of the phase noise difference $\delta_\phi(n,l)$ [8], the 3 dB bandwidth $f_{3\,\text{dB}}$ determines the quality of the oscillator and α_l is the power of the lth path.

In the second and third architectures, the phase noise-induced SI is partially reduced, depending on the accuracy of the estimated SI channel. In this case, p_{pn} can be expressed as:

$$p_{\text{pn}} = p_{\text{si,Tx}} \left(\frac{2}{N_{\text{training}}} \sum_{l=0}^{L-1} \alpha_l \left(1 - e^{-\frac{4\pi f_{3\,\text{dB}} T_s l}{2}}\right) + \frac{\sigma^2}{p_{\text{si,in}} N_{\text{training}}} \right) \sum_{l=0}^{L_{\text{RF}}-1} \sigma^2_{\text{pn},l} + \sum_{l=L_{\text{RF}}}^{L-1} \sigma^2_{\text{pn},l} \alpha_l p_{\text{si,Tx}}, \tag{3.14}$$

where we used the expression of the channel estimation error provided in the Appendix for the particular case of the LS estimator and N_{training} is the number of samples in the initial half-duplex period.

3.1.3 *Quantization and Clipping Noise from ADC*

Most of the communications systems use uniform ADCs with 2^b equidistant quantization levels, where b is the number of bits of the ADC. Assuming that the output signal from the ADC is limited between $+1$ and -1, the quantization function is given by [9]:

$$Q(x)=\begin{cases}-1, & \text{if } x\leq\frac{2-2^b}{2^b-1},\\ 2\frac{q-1}{2^b-1}-1, & \text{if } \frac{2q-2-2^b}{2^b-1}<x\leq\frac{2q-2^b}{2^b-1},\ q=2,\ldots,\ 2^b-1,\\ 1, & \text{if } x\geq\frac{2^b-2}{2^b-1}.\end{cases}$$

In practice, the ADC is preceded by a VGA to fit the input signal to the operating range of the ADC. As a result, the presence of the SI decreases the amount of bits used for the intended signal.

The classical relation between the signal and the quantization noise $6.02b + 1.76\,\text{dB}$ was derived assuming a sinusoidal input signal [10]. When using OFDM modulation, the resulting signal is well approximated by a Gaussian process due to the central limit theorem [11]. Thus, the formula $6.02b + 1.76$ is no longer valid. Using the Bussgang's theorem for nonlinear memoryless systems [12], the output samples of the ADC are an attenuated version of the input plus a statistical independent Gaussian term. Thus the input-output relation of the ADC is given by:

$$y_{\text{BB}}(n) = \alpha\sqrt{g_{\text{VGA}}}y(n) + n_{\text{ADC}}(n), \tag{3.15}$$

where $\{y(n)\}$ are samples of the input signal to the ADC, α is expressed as:

$$\alpha = \sqrt{g_{\text{VGA}}}\frac{\mathbb{E}\{y^*(n)y_{\text{BB}}(n)\}}{\mathbb{E}\{|y(n)|^2\}}, \tag{3.16}$$

and $n_{\text{ADC}}(n)$ is the additive clipping-plus-quantization noise. The power of n_{ADC} is obtained by rearranging the terms in (3.15) as:

$$p_{\text{ADC}} = \mathbb{E}\{|y_{\text{BB}}(n)|^2 - \alpha^2 g_{\text{VGA}}\mathbb{E}\{|y(n)|^2\}. \tag{3.17}$$

The various expectations in (3.16) and (3.17) can be easily evaluated using the Gaussian distribution of the input signal:

$$\begin{aligned}\mathbb{E}\{|y_{\text{BB}}(n)|^2\} &= \int_{-\infty}^{+\infty} Q^2(x)f(x, g_{\text{VGA}}p_{\text{VGA,in}})dx,\\ \mathbb{E}\{y^*(n)y_{\text{BB}}(n)\} &= \int_{-\infty}^{+\infty} \frac{xQ(x)}{\sqrt{g_{\text{VGA}}}}f(x, g_{\text{VGA}}p_{\text{VGA,in}})dx,\end{aligned} \tag{3.18}$$

$p_{\text{VGA,in}}$ is the power of the signal at the input of the VGA and $f(x, \sigma_x^2)$ is the probability distribution function of the Gaussian variable x with zero mean and variance σ_x^2.

3.1.4 Transceiver Impairments and Noise

While it is relatively easy to reduce the linear part of the SI, reducing the different impairments from the Tx and Rx chains is more challenging, especially when these impairments are not correlated with the known transmit signal. In the following, we evaluate the power of the impairments at the input of the ADC and expressed in (3.7) and (3.11). This is an important part as the total power of the SI dictates the power of the quantization noise from the ADC.

The additive Gaussian noise represents the thermal noise inherent in the transceiver circuits and is usually characterized by the noise factor. The output noise power from the receiver when considering the multiple stage in Fig. 3.1 is:

$$p_{\text{noise,Rx}} = kT_0\, F_{\text{Rx}} BW(g_{\text{LNA}} g_{\text{IQ}}), \tag{3.19}$$

where g_{LNA} and g_{IQ} are the power gains of the LNA and the IQ mixer, BW is the bandwidth of the signal, the thermal noise power spectral density is $kT_0 = 10^{-174/10}$ mW/Hz and F_{Rx} is the overall noise factor of the receiver which can be calculated using the Friis' formula from the individual noise factors F_{LNA}, F_{IQ} and F_{VGA} of the LNA, IQ mixer and VGA, respectively, as:

$$F_{\text{Rx}} = F_{\text{LNA}} + \frac{F_{\text{IQ}} - 1}{g_{\text{LNA}}} + \frac{F_{\text{VGA}} - 1}{g_{\text{LNA}} g_{\text{IQ}}}. \tag{3.20}$$

The Tx noise emission in the same receiver band affects also the performance of full-duplex systems. While this noise is partially reduced in the second and third architectures, it is not the case in the first architecture, and has also to be derived to properly evaluate the performance of the overall system. Thus, denoting by $F_{\text{Tx}} = F_{\text{IQ}} + \frac{F_{\text{PA}-1}}{g_{\text{IQ}}}$ the noise factor of the Tx chain, the noise emission from the transmitter $p_{\text{noise,Tx}}$ is given by:

$$p_{\text{noise,Tx}} = kT_0\, BW\, g_{\text{Tx}} F_{\text{Tx}}, \tag{3.21}$$

where $g_{\text{Tx}} = g_{\text{IQ}} g_{\text{PA}}$ is the gain of the transmitter and $\{F_{\text{PA}}, g_{\text{PA}}\}$ and $\{F_{\text{IQ}}, g_{\text{IQ}}\}$ are the {noise factor, gain} of the PA and the transmit IQ mixer, respectively. In the previous expressions, it is assumed that the noises generated by other devices, such as lowpass and bandpass filters, are negligibly low compared to the noise generated by the PA and IQ mixer.

Besides the additive noise, the nonlinear distortions produced by the Tx and Rx chains also have to be modeled. The IQ mixer creates an inband image of the signal [13]. The power of the inband image depends on the image rejection capability α_{IQ} of the IQ mixer as:

$$p_{\text{image}} = \alpha_{\text{IQ}} p_{\text{in}}, \tag{3.22}$$

where p_{in} is the signal power at the input of the IQ mixer. For the Tx PA and the receive LNA, the nth order nonlinearity is related to the power of the input signal p_{in} as [4]:

$$p_{n,\mathrm{nl}} = \frac{p_{\mathrm{in}}^n g}{iip_n^{n-1}}, \tag{3.23}$$

where iip_n is the nth order input reference intercept point and g is the linear gain of the component. Such nonlinearity model has the major advantage of making the following analysis possible.

In the Tx chain, the main nonlinearities are generated by the IQ mixer and the PA because of its large gain. We assume that the PA produces nonlinear distortions up to order $2P+1$. Using (3.22) and (3.23), the accumulated power of the essential distortions from the transmitter is given by:

$$p_{\mathrm{nl,Tx}} = g_{\mathrm{PA}}\alpha_{\mathrm{IQ}}p_{\mathrm{in}} + \sum_{n=1}^{P} \frac{(g_{\mathrm{IQ}}+\alpha_{\mathrm{IQ}})^{2n+1} p_{\mathrm{in}}^{2n+1} g_{\mathrm{PA}}}{iip_{2n+1}^{2n}}, \tag{3.24}$$

where only the odd orders falling in the signal band are considered. This distortion can be reduced by the RF cancellation in the second and third architectures since it is already included in the reference signal. In contrast, in the first architecture, it is not included in the reference signal for RF cancellation which increases the SI power at the ADC input, and the quantization noise. On the other hand, nonlinear baseband cancellation can be used to reduce the PA-induced nonlinearity in the first architecture, leaving the quantization noise as the main limitation of this architecture. The limitations of the three architectures will be further discussed in Sect. 3.2.

Similarly, the Rx chain also introduces inband nonlinearities, mainly from the LNA and the IQ mixer. Thus the total power of the Rx-induced distortions can be written as:

$$p_{\mathrm{nl,Rx}} = \alpha_{\mathrm{IQ}} g_{\mathrm{LNA}} p_{\mathrm{total}} + (g_{\mathrm{IQ}}+\alpha_{\mathrm{IQ}}) \sum_{n=1}^{P} \frac{p_{\mathrm{total}}^{2n+1} g_{\mathrm{LNA}}}{iip_{2n+1}^{2n}} \tag{3.25}$$

where p_{total} is the total power of the received signal containing the intended signal, the additive noise and the SI after RF cancellation.

3.1.5 Auxiliary Component-Induced Impairments

This section discusses the effects of the additional components used in the first architecture for up-conversion of the baseband signal for the RF cancellation stage and in the third architecture for down-conversion of the PA output for the baseband cancellation stage.

The first architecture requires an auxiliary Tx, including an additional set of DAC, IQ mixer and PA. The accumulated power of the resulting distortions is expressed in a similar way to the Tx distortion in (3.24). However, the auxiliary PA needs lower power and smaller gain than the PA used in the main Tx chain. Therefore, its resulting impairment power is lower than the main Tx-induced impairment power. On the other hand, the third architecture uses the transmitted signal after the PA as the reference signal for the baseband cancellation. This requires an auxiliary Rx chain containing an additional set of IQ mixer and ADC.

3.1.6 Baseband SI-Cancellation

In the first and second architectures, nonlinear impairments affecting the received SI and the SI channel after the RF cancellation stage can be estimated and used with the baseband reference signal to construct. The SI-replica for baseband SI-cancellation. In this case, the power of the residual SI after the baseband SI-cancellation is given by:

$$p_{\mathrm{si,BB}} = \left(\frac{p_{\mathrm{si,in}}}{\beta_{\mathrm{RF}}\beta_{\mathrm{BB}}} + \frac{p_{\mathrm{nl,Tx}}}{\beta_{\mathrm{BB}}} + p_{\mathrm{pn}} + p_{\mathrm{nl,aux,Tx}} + p_{\mathrm{noise,Tx}} \right) g_{\mathrm{Rx}} + p_{\mathrm{nl,Rx}}, \tag{3.26}$$

for the first architecture, and by:

$$p_{\mathrm{si,BB}} = \left(\frac{p_{\mathrm{si,in}}^{(1)} + p_{\mathrm{nl,Tx}}}{\beta_{\mathrm{RF}}\beta_{\mathrm{BB}}} + \frac{p_{\mathrm{noise,Tx}}}{\beta_{\mathrm{RF}}} + \frac{p_{\mathrm{si,in}}^{(2)}}{\beta_{\mathrm{BB}}} + p_{\mathrm{pn}} \right) g_{\mathrm{Rx}} + p_{\mathrm{nl,Rx}}, \tag{3.27}$$

for the second architecture where β_{BB} is the amount of suppression achieved by the baseband SI-cancellation stage. The difference in the residual SI power given in (3.26) and (3.27) comes essentially from the RF SI-cancellation stage operation, i.e., the second architecture uses the PA output as reference signal and models some paths of the SI channel, while the first architecture can model all the paths to cancel in the RF stage. The impairments resulting from the additional up-conversion in the first architecture are also included in the residual SI.

On the other hand, the third architecture uses only linear baseband cancellation by convolving the reference signal with the estimated SI channel without using non-linear cancellation since the reference signal already contains the Tx impairments. The resulting residual SI power is expressed as:

$$p_{\mathrm{si,BB}} = \left(\frac{p_{\mathrm{si,in}}^{(1)} + p_{\mathrm{nl,Tx}} + p_{\mathrm{noise,Tx}}}{\beta_{\mathrm{RF}}\beta_{\mathrm{BB}}} + \frac{p_{\mathrm{si,in}}^{(2)}}{\beta_{\mathrm{BB}}} + p_{\mathrm{pn}} \right) g_{\mathrm{Rx}} + p_{\mathrm{nl,Rx}} + p_{\mathrm{nl,aux,Rx}}, \tag{3.28}$$

which is similar to the residual SI obtained for the second architecture with the exception that the Tx noise is reduced in the baseband and the presence of the additional term $p_{\mathrm{nl,aux,Rx}}$ representing the distortions generated by the auxiliary down-conversion chain, used to obtain the baseband reference signal.

3.2 Performance Evaluation: Illustrative Results

In this section, using the derivations in the previous section, we evaluate the impacts of all components on the overall system performance in the three architectures. Let SIR and SNR denote the intended Rx signal-to-interference power ratio before any cancellation and the intended Rx signal-to-noise power ratio, respectively. In the following sub-sections, for illustrative examples, we assume SIR$= -100$ dB (i.e., the intended Rx signal is 100 dB lower than the SI due to propagation loss).

3.2.1 System Specifications

3.2.1.1 SI Channel Modelling

The characterisation of the SI channel is an important part toward the analysis as the RF stage for the second and third architectures depends on both the number of paths and the delay profile of the channel. Various measurements have been done to characterise the SI channel. Consider the simple and popular architecture using the same antenna to transmit and receive via a 3-port circulator, the dominant paths of the SI channel come from the leakage through the circulator and the internal antenna reflections due to the impedance mismatch between the isolator and the antenna. On the other hand, external reflections from closely-located objects may occur with much larger delays and weaker levels compared to the dominant paths since they travel longer distances. It was reported in [14] that the external reflections are about 30 dB lower than the leakage and antenna reflection paths. When using two different antennas to transmit and receive, the line-of-sight (LoS) components and the path coming from the electromagnetic waves reflected from the transceiver structure represent the most significant paths [15, 16]. In both cases, the internal reflections are quasi-static since they depend on the transceiver structure while the external reflections vary according to the surrounding environment. In general, the power delay profile (PDP) of the SI channel can be written as [14]:

$$\mathrm{PDP}(t,\tau) = \alpha_{m_1}\delta(t-\tau_{m_1}) + \alpha_{m_2}\delta(t-\tau_{m_2}) + \sum_{l=2}^{L-1}\alpha_l\delta(t-\tau_l), \tag{3.29}$$

Table 3.1 Example transceiver parameters

Parameter	Value
PA OIP3	47 dBm
PA 1 dB Output compression point	37 dBm
PA noise figure[a]	4 dB
IQ mixer gain (in the Tx and Rx chains)	6 dB
IQ mixer image rejection	30 dB
IQ mixer noise figure	4 dB
LNA gain	25 dB
LNA IIP3	−8 dBm
LNA noise figure	4 dB
VGA noise figure	10 dB
ADC	12 bits
N_{training}	1000
Bandwidth	20 MHz
Thermal noise power	−101 dBm
Intended Rx signal power	−81 dBm
Tx-Rx antenna isolation	40 dB

[a]The noise figure is equivalent to the noise factor expressed in dB scale

where $(\alpha_{m_1}, \tau_{m_1})$ and $(\alpha_{m_2}, \tau_{m_2})$ are the power/delay of the internal/coupling reflections and (α_l, τ_l), for $l = 2, \ldots, L-1$ are the power/delay of the external reflections. In the following, we consider $L = 9$.

3.2.1.2 Transceiver Parameters

For illustrative results, we assume a signal bandwidth of 20 MHz and a typical Tx-Rx isolation of 40 dB [17, 18] and other typical transceiver parameters [4, 6, 7, 19, 20] as summarized in Table 3.1.

3.2.2 *Residual Self-Interference After RF Cancellation Stage*

The RF cancellation stage subtracts a replica of the received SI, based on an estimate of the SI channel, from the received RF signal. As discussed in Sect. 3.1, all the channel paths can be cancelled using the first architecture while a limited number of paths are cancelled in the other two architectures. The performance of the RF cancellation stage highly depends on the SI channel estimation accuracy, and the number of modeled paths for the second and third architectures. We suppose that the SI channel is estimated during an initial half-duplex transmission period (i.e., without the intended signal) using the LS estimator. In this case, the amount of RF cancellation β_{RF} for the first architecture is given by (more details on the derivations can be found in the Appendix):

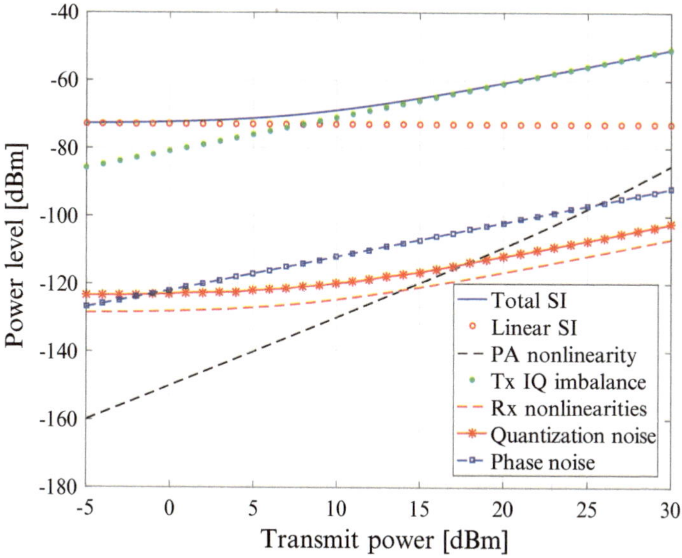

Fig. 3.2 Power levels of the different components after the ADC versus the transmit power for the first architecture

$$\beta_{\mathrm{RF}} = \frac{p_{\mathrm{si,Tx}} N_{\mathrm{training}}}{2Lp_{\mathrm{si,Tx}} \sum_{l=0}^{L-1} \alpha_l \left(1 - e^{-\frac{4\pi f_{3\mathrm{dB}} T_s l}{2}}\right) + L\sigma^2}. \tag{3.30}$$

The LS estimator is used here to obtain insight into the achievable RF cancellation. As discussed in the previous section, all paths of the SI channel are cancelled in the first architecture while leaving the Tx impairments to be reduced in the baseband. On the other hand, the second and third architectures reduce a selected number of paths, along with the Tx impairments, in the RF stage. Thus we cannot intuitively determine which architecture presents better RF cancellation performance. To answer this question, we study and plot the power levels computed at the ADC output of the individual components after RF SI-cancellation versus the transmit power in Fig. 3.2 for the first architecture (using (3.7), (3.13), (3.17) and the detailed expressions in Sect. 3.1.4) and in Fig. 3.3 for the second and third architectures (using (3.11), (3.14), (3.17) and the detailed expressions in Sect. 3.1.4) using $N_{\mathrm{training}} = 1000$. We consider $L_{\mathrm{RF}} = 2$ to cancel the main two paths. For the first architecture, the most significant residual SI comes from the Tx impairments, which cannot be reduced by the RF SI-cancellation stage while the linear part of the SI represents the main part of the residual SI in the other two architectures as only a limited number of paths are cancelled.

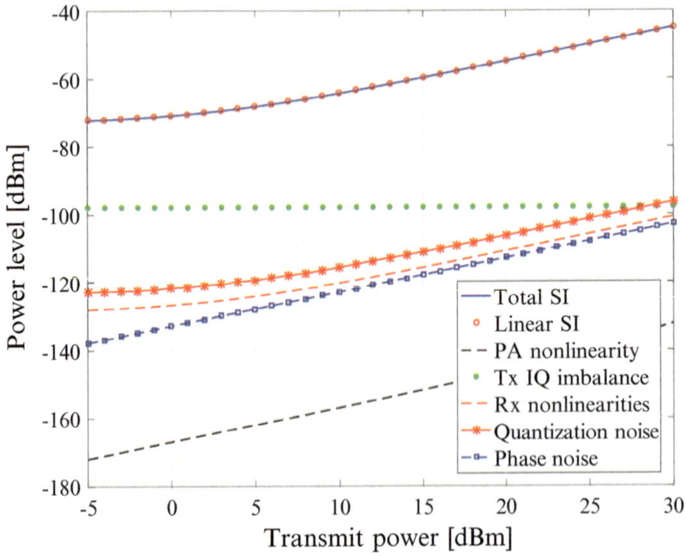

Fig. 3.3 Power levels of the different components after the ADC using the second or third architectures

3.2.2.1 Quantization Noise

To better illustrate the effects of the ADC quantization noise, Fig. 3.4 plots the intended signal, the residual SI and the thermal noise to the quantization noise. In Fig. 3.4, we distinguish between two regions depending on whether the thermal noise is larger or smaller than the quantization noise. When the transmit power is higher than 25 dBm for the first architecture or 20 dBm for the second and third architectures, the intended signal-to-thermal-plus-quantization noise ratio is more than 1 dB lower than the intended signal-to-noise power ratio (SNR), as the quantization noise becomes dominant.

Therefore, the RF cancellation stage should guarantee that the quantization noise remains lower than the thermal noise. For example, to keep a degradation less than 1 dB (i.e., $10\log_{10}(p_{\text{thermal}}+p_{\text{ADC}})-10\log_{10}(p_{\text{thermal}}) < 1$ dB), the quantization noise should satisfy:

$$10\log_{10}(p_{\text{thermal}}) - 10\log_{10}(p_{\text{ADC}}) > -10\log_{10}(10^{0.1} - 1) = 5.86\,\text{dB}. \qquad (3.31)$$

Figure 3.5 shows the minimum amount of RF cancellation required to keep the quantization noise satisfying the condition in (3.31) for different numbers of bits in the ADC. The curves with square markers are for the first architecture and the other curves are for the second and third architectures. Clearly, as the number of bits decreases, the required amount of RF cancellation increases to keep the quantization noise below the limit. For high transmit power and a low number of ADC bits,

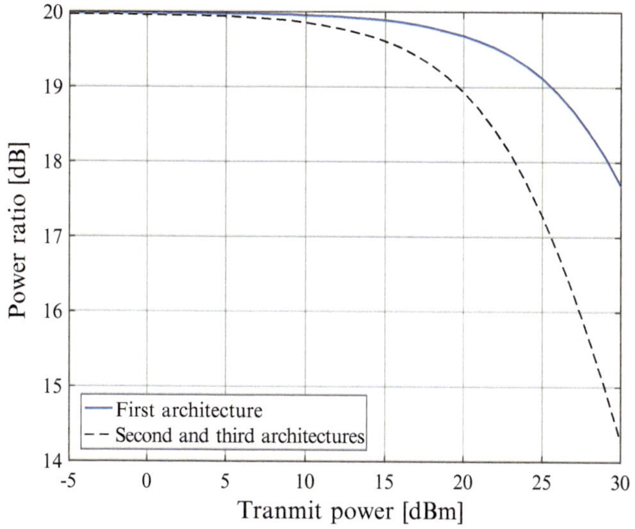

Fig. 3.4 Intended signal-to-thermal-plus-quantization-noise power ratio versus transmit power

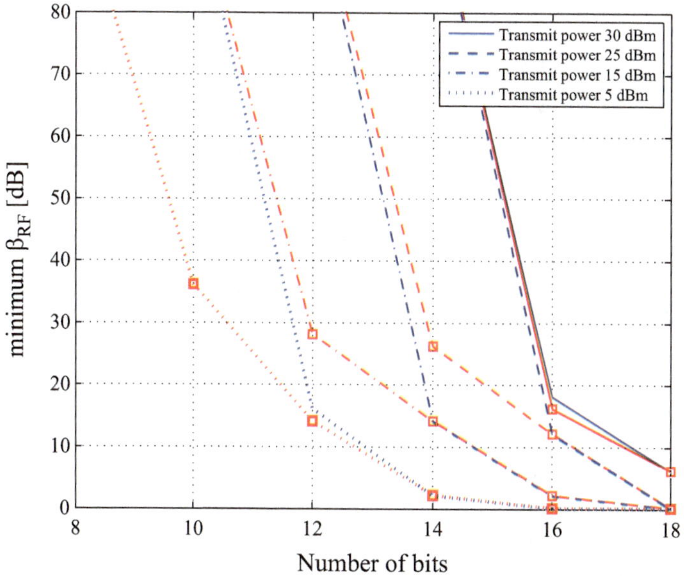

Fig. 3.5 Minimum required β_{RF} versus number of bits in the ADC to keep degradation in intended signal-to-thermal-plus-quantization noise ratio less than 1 dB

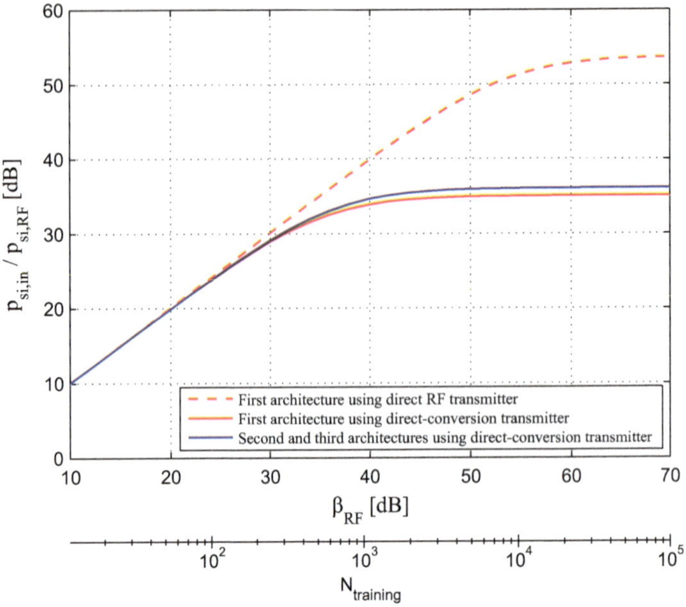

Fig. 3.6 Power ratio of the SI before and after the RF cancellation stage versus β_{RF}, for transmitting power for 30 dBm

the quantization noise becomes a critical factor as the required RF cancellation increases to infinity indicating that either the Tx impairments in the first architecture or the non-cancelled paths in the second and third architectures become the limiting factors.

3.2.2.2 Limiting Behaviour of the RF Cancellation Stage

From (3.7) and (3.11), the RF cancellation stage reduces the linear SI in the first architecture and both the linear and Tx impairments coming from the modeled paths in the second and third architectures by β_{RF} that linearly increases with increasing training length N_{training}. We define the *RF SI-cancellation gain* as the power ratio of the SI before and after the RF cancellation (i.e., $p_{\text{si,in}}/p_{\text{si,RF}}$), and plot it versus β_{RF} (with the corresponding training length N_{training} according to (3.30)) in Fig. 3.6. We also included, for comparison, the RF SI-cancellation gain $p_{\text{si,in}}/p_{\text{si,RF}}$ when using the direct RF Tx. Direct RF Tx performs up-conversion and IQ mixer in the digital domain which reduces considerably the IQ imbalance compared to direct-conversion Tx. As β_{RF} increases, the RF SI-cancellation gain for the first architecture saturates at 35 dB, where a similar number was reported in [21, 22] from experimental measurements. This limitation results from the Tx impairments where we can see from Fig. 3.2 that the signal image from the transmit IQ mixer is indeed the limiting factor for the RF cancellation stage. Therefore, direct RF Tx can offer higher cancellation gain as the signal image can be negligible. When using the second and third architectures, the RF SI-cancellation gain is limited by the relative level of the reflection paths compared to the two main paths. The direct RF Tx is not represented for these architectures since it presents the same performance as the direct-conversion Tx.

3.2.3 Residual SI After the Baseband SI-Cancellation Stage

Prior to the detection, the residual SI is reduced in the baseband by subtracting a baseband replica of the residual SI from the received signal. In the first and second architectures, the SI-replica samples are generated by processing the known transmitted baseband symbols with an estimate of the nonlinearity coefficients and the residual SI channel after the RF cancellation stage. The estimated residual SI channel includes the effects of the transceiver and the multipath components due to the external reflections. When the third architecture is considered, the reference signal for baseband cancellation is taken from the output of the PA. This means that the Tx nonlinearities are already included on the reference signal and only linear processing is needed to create the SI-replica for cancellation. To analyze the cancellation limits, we assume the amount of baseband suppression β_{BB} to be arbitrary increased. With this assumption, it is possible to determine the best performance that can be obtained by the baseband cancellation stage. Figure 3.7 plots the *baseband SI – cancellation gain* (i.e., the power ratio of the SI before and after the baseband cancellation $p_{\mathrm{si,RF}}/p_{\mathrm{si,BB}}$) versus β_{BB} where β_{RF} follows (3.30). It can be seen that the ratio $p_{\mathrm{si,RF}}/p_{\mathrm{si,BB}}$ first increases dB-by-dB with β_{BB} and then saturates as β_{BB} gets larger than some level. One interesting observation is that, despite that the same baseband cancellation stage is implemented in the first and second architectures, their performance is different.

In the following, we further investigate the behavior of the curves in Fig. 3.7. We analyze the power of the individual signal components expressed in (3.26), (3.27) and (3.28) after the baseband cancellation, represented in Figs. 3.8, 3.9 and 3.10 for the three architectures. First, same phenomena appears for the three architectures. The residual SI due to phase noise limits the cancellation performance of the full-duplex transceiver. This residual SI and the quantization noise cannot be cancelled in the baseband stage. Actually, the phase noise is a time-varying process, making its compensation challenging, contrarily to the SI channel and Tx nonlinearities that can be modeled using constant coefficients over a given time interval. However, some methods have been proposed to estimate and cancel the phase noise effect [23, 24]. On the other hand, the quantization noise is completely random and uncorrelated with the transmitted SI, making its cancellation almost impossible. Another observation from Figs. 3.7, 3.8, 3.9 and 3.10 is that the RF cancellation stage dictates the performance limit of the baseband cancellation stage. Although the first and second architectures are based on the same baseband cancellation

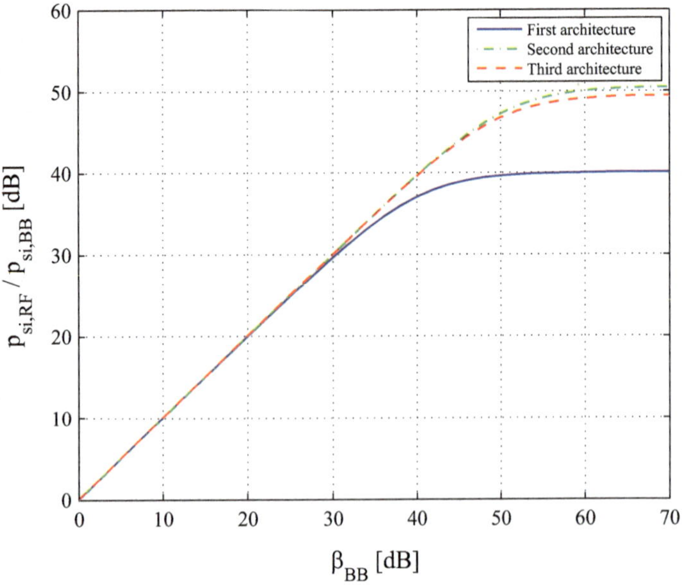

Fig. 3.7 Power ratio of the SI before and after the baseband cancellation stage versus β_{BB}, for transmitting power of 30 dBm

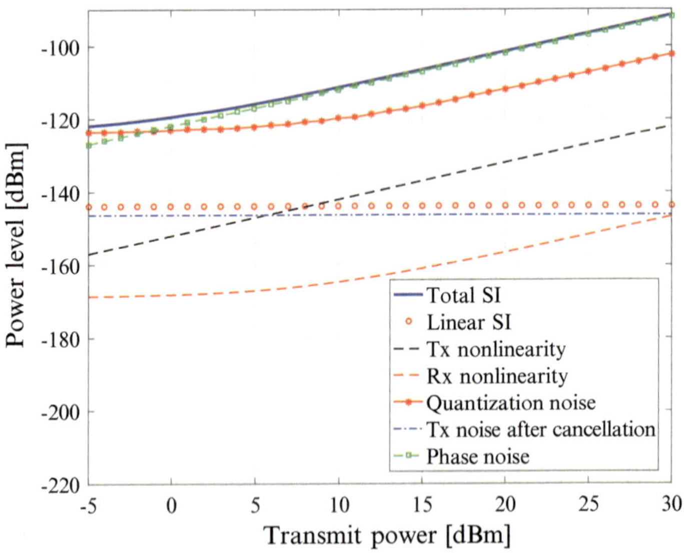

Fig. 3.8 Power levels of the different components after the baseband cancellation stage using the first architecture

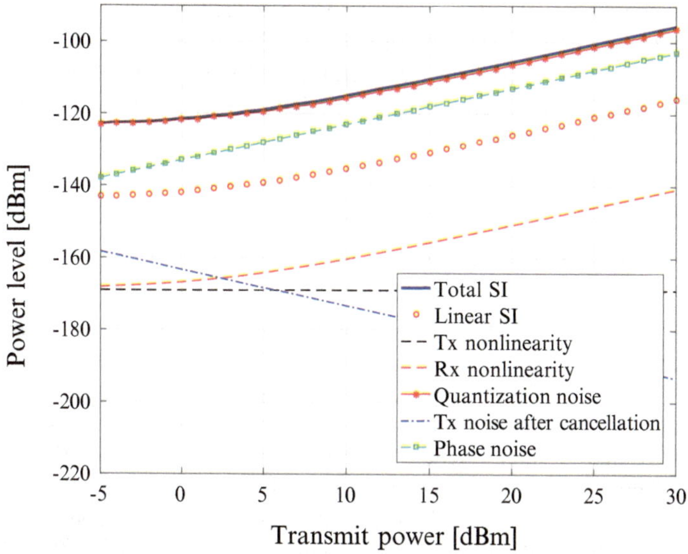

Fig. 3.9 Power levels of the different components after the baseband cancellation stage using the second architecture

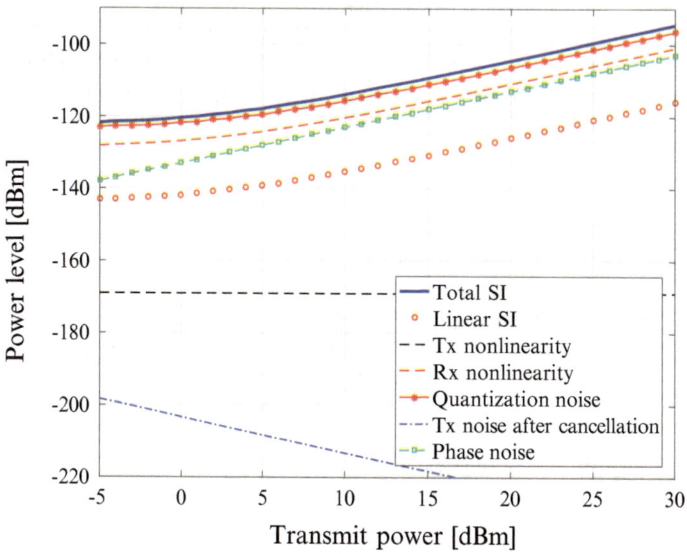

Fig. 3.10 Power levels of the different components after the baseband cancellation stage using the third architecture

procedure, the second architecture offers better cancellation performance than the first architecture. As the phase noise-induced SI is partially cancelled in the RF cancellation stage of the second architecture, the following baseband cancellation stage has more margins to reduce the SI, leading to better cancellation performance than the RF cancellation stage.

3.3 Conclusion

Analysing the SI is one of the most important aspects to study towards developing full-duplex systems. By covering the three widely used architectures for SI-cancellation, we identified the main limitations for every cancellation stage. It turns out that the RF cancellation stage, when taking the baseband signal as a reference, is limited by the Tx impairments. On the other hand, using the PA output as a reference signal for the RF cancellation stage has the advantage of reducing the Tx impairments as well. The performance of this method depends on the number of cancelled paths. It was also observed that the phase noise can be partially reduced for the modeled paths in the RF cancellation stage. In this case, the maximum amount of total cancellation, from the RF and the baseband cancellation stages, is limited by the quantization noise of the receiver ADC. When taking the baseband signal as reference for the RF cancellation stage, the phase noise effect is not reduced, making it the main limiting factor. It was found that the phase noise is not the major bottleneck for the cancellation performance with transmit power below 20 dBm, bringing the residual SI below the thermal noise. Also, the Tx nonlinearities have to be reduced in the baseband.

Appendix

First, we consider a training period used to estimate the SI channel for the RF cancellation stage, where the transceiver receives only its transmitted signal. In this case, the received signal in the baseband is given by:

$$\begin{aligned} y(n) &= \sum_{l=0}^{L-1} h^{\text{si}}(l)z(l)e^{j\delta_\phi(n,l)} + n(n) \\ &\overset{(a)}{=} \sum_{l=0}^{L-1} h^{\text{si}}(l)z(l) + (e^{j\delta_\phi(n,l)} - 1)h^{\text{si}}(l)z(l) + n(n). \end{aligned} \tag{3.32}$$

The reason for the formulation in $\overset{(a)}{=}$ is to highlight the effect of the phase noise on the estimation SI channel estimation performance since it is considered as additive noise included in the term n_{total}. It can be verified that the power of n_{total} is equal to:

$$p_{n_{\text{total}}} = 2p_{\text{si,Tx}} \sum_{l=0}^{L-1} \alpha_l \left(1 - e^{-\frac{w_c^2 c T_s l}{2}}\right) + \sigma^2, \tag{3.33}$$

which follows from the characteristic function of the free-running oscillator [25]. One approach of channel estimation is the least-square method, which estimates the SI channel from a set of N received samples $\boldsymbol{y} = [y(1), \ldots, y(N)]^T$ as:

$$\widehat{\boldsymbol{h}}^{\text{si}} = \left(\boldsymbol{Z}^H \boldsymbol{Z}\right)^{-1} \boldsymbol{Z}^H \boldsymbol{y}, \tag{3.34}$$

where $\boldsymbol{Z}$ is a Toeplitz matrix collecting the known transmit signal and $\widehat{\boldsymbol{h}}^{si} = \left[\widehat{h}^{\text{si}}(0), \ldots, \widehat{h}^{\text{si}}(L-1)\right]^T$ is the vector gathering the estimated SI channel. The corresponding mean square error (MSE) of the estimated SI channel is given by:

$$\begin{aligned} \text{MSE} &= \text{trace}\, \mathbb{E}\left\{(\boldsymbol{h}^{\text{si}} - \widehat{\boldsymbol{h}}^{\text{si}})(\boldsymbol{h}^{\text{si}} - \widehat{\boldsymbol{h}}^{\text{si}})^H\right\} \\ &= \frac{2Lp_{\text{si,Tx}} \sum_{l=0}^{L-1} \alpha_l \left(1 - e^{-\frac{w_c^2 c T_s l}{2}}\right) + L\sigma^2}{p_{\text{si,Tx}} N_{\text{training}}}. \end{aligned} \tag{3.35}$$

Using the estimated $\widehat{\boldsymbol{h}}^{\text{si}}$ for the RF cancellation stage, in the first architecture, the amount of cancellation α_{RF} can be obtained as:

$$\alpha_{\text{RF}} = \frac{p_{\text{si,in}}}{\text{trace}\left\{\mathbb{E}\left\{\boldsymbol{Z}(\boldsymbol{h}^{\text{si}} - \widehat{\boldsymbol{h}}^{\text{si}})(\boldsymbol{h}^{\text{si}} - \widehat{\boldsymbol{h}}^{\text{si}})^H \boldsymbol{Z}^H\right\}\right\}} \tag{3.36}$$

which gives the expression in (3.30).

References

1. D. Korpi, T. Riihonen, V. Syrjala, L. Anttila, M. Valkama, and R. Wichman, "Full-duplex transceiver system calculations: Analysis of ADC and linearity challenges," *IEEE Trans. Wireless Comm.*, vol. 13, no. 7, pp. 3821–3836, 2014.
2. B. P. Day, A. R. Margetts, D. W. Bliss, and P. Schniter, "Full-duplex bidirectional MIMO: Achievable rates under limited dynamic range," *IEEE Trans. Signal Process*, vol. 60, no. 7, pp. 3702–3713, 2012.
3. V. Syrjala, M. Valkama, L. Anttila, T. Riihonen, and D. Korpi, "Analysis of oscillator phase-noise effects on self-interference cancellation in full-duplex OFDM radio transceivers," *IEEE Trans. on Wireless Comm.*, vol. 13, no. 6, pp. 2977–2990, June 2014.
4. Q. Gu, *RF system design of transceivers for wireless communications*. Springer Science & Business Media, 2005.
5. A. Parssinen, J. Jussila, J. Ryynänen, L. Sumanen, and K. A. Halonen, "A 2-GHz wide-band direct conversion receiver for WCDMA applications," *IEEE Journal of Solid-State Circuits*, vol. 34, no. 12, pp. 1893–1903, 1999.

6. H. Yoshida, T. Kato, T. Toyoda, I. Seto, R. Fujimoto, T. Kimura, O. Watanabe, T. Arai, T. Itakura, and H. Tsurumi, "Fully differential direct conversion receiver for W-CDMA using an active harmonic mixer," in *IEEE Radio Frequency Integrated Circuits (RFIC) Symposium*, June 2003.
7. B. Razavi and R. Behzad, *RF microelectronics*. Prentice Hall New Jersey, 1998, vol. 2.
8. A. Demir, A. Mehrotra, and J. Roychowdhury, "Phase noise in oscillators: a unifying theory and numerical methods for characterization," *IEEE Trans. on Circuits and Syst.*, vol. 47, no. 5, pp. 655–674, 2000.
9. A. Masmoudi and T. Le-Ngoc, "Residual self-interference after cancellation in full-duplex systems," in *Proc. IEEE Int. Conf. Commun.*, Sydney, Australia, June 2014.
10. A. V. Oppenheim, R. W. Schafer, J. R. Buck *et al.*, *Discrete-time signal processing*. Prentice hall Englewood Cliffs, NJ, 1989, vol. 2.
11. S. Wei, D. L. Goeckel, and P. A. Kelly, "Convergence of the complex envelope of bandlimited OFDM signals," *IEEE Trans. Inform. Theory*, vol. 56, no. 10, pp. 4893–4904, 2010.
12. H. Rowe, "Memoryless nonlinearities with Gaussian inputs: Elementary results," *Bell System Technical Journal*, vol. 61, no. 7, pp. 1519–1526, 1982.
13. L. Anttila, "Digital front-end signal processing with widely-linear signal models in radio devices," *Tampereen teknillinen yliopisto. Julkaisu-Tampere University of Technology. Publication; 985*, 2011.
14. X. Wu, Y. Shen, and Y. Tang, "The power delay profile of the single-antenna full-duplex self-interference channel in indoor environments at 2.6 GHz," *IEEE Antennas and Wireless Propagation Letters*, vol. 13, pp. 1561–1564, 2014.
15. C. R. Anderson, S. Krishnamoorthy, C. G. Ranson, T. J. Lemon, W. G. Newhall, T. Kummetz, and J. H. Reed, "Antenna isolation, wideband multipath propagation measurements, and interference mitigation for on-frequency repeaters," in *Proc. IEEE SoutheastCon*, Greensboro, North Carolina, USA, Mar. 2004.
16. A. Sethi, V. Tapio, and M. Juntti, "Self-interference channel for full duplex transceivers," in *Proc. IEEE Wireless Commun. and Netw. Conf.*, Istanbul, Turkey, Apr. 2014.
17. E. Everett, A. Sahai, and A. Sabharwal, "Passive self-interference suppression for full-duplex infrastructure nodes," *IEEE Trans. Wireless Comm.*, vol. 13, no. 2, pp. 680–694, 2014.
18. J. I. Choi, M. Jain, K. Srinivasan, P. Levis, and S. Katti, "Achieving single channel, full duplex wireless communication," in *Proc. ACM MobiCom*, Chicago, Illinois, USA, Sept. 2010.
19. A. Masmoudi and T. Le-Ngoc, "Self-interference cancellation limits in full-duplex communication systems," in *Proc. IEEE Global Telecommun. Conf.*, Washington DC, USA, Dec. 2016.
20. A. Behzad, *Wireless LAN radios: System definition to transistor design*. John Wiley & Sons, 2007.
21. M. Duarte, C. Dick, and A. Sabharwal, "Experiment-driven characterization of full-duplex wireless systems," *IEEE Trans. Wireless Comm.*, vol. 11, no. 12, pp. 4296–4307, 2012.
22. M. Duarte and A. Sabharwal, "Full-duplex wireless communications using off-the-shelf radios: Feasibility and first results," in *Proc. ASILOMAR Signals, Syst., Comput.*, Pacific Frove, California, USA, Nov. 2010.
23. A. Masmoudi and T. Le-Ngoc, "A maximum-likelihood channel estimator for self-interference cancellation in full-duplex systems," *IEEE Trans. Vehicular Technology*, vol. 65, no. 7, pp. 5122–5132, July 2016.
24. R. Li, A. Masmoudi, and T. Le-Ngoc, "Self-interference cancellation with phase-noise suppression in full-duplex systems," in *Proc. IEEE Pers. Indoor and Mobile Radio Commun.*, Hong Kong, China, Aug. 2015.
25. D. Petrovic, W. Rave, and G. Fettweis, "Effects of phase noise on OFDM systems with and without PLL: Characterization and compensation," *IEEE Trans. on Comm.*, vol. 55, no. 8, pp. 1607–1616, 2007.

Chapter 4
Self-Interference Channel Estimation and Cancellation Using Compressed-Sensing and Subspace Approaches

Abstract This chapter presents a two-stage self-interference (SI) cancellation for full-duplex MIMO communications systems. By exploiting the SI channel sparsity, a compressed-sensing based SI channel estimation technique is developed and used in the first SI cancellation radio-frequency (RF) stage to reduce the SI power prior to the receiver low-noise amplifier (LNA) and the analog-to-digital converter (ADC) to avoid overloading. Subsequently, a subspace-based algorithm is proposed to jointly estimate the coefficients of both the residual SI and intended channels, and transceiver impairments for the second SI cancellation baseband stage to further reduce the residual SI. Unlike other previous works, the intended signal is taken into consideration during the estimation process to reduce the overhead. It is demonstrated that the SI channel coefficients can be perfectly estimated without any knowledge of the intended signal, and only a few training symbols are needed for ambiguity removal in intended-channel estimation. Simulation results show that the proposed algorithms outperform the Least Square (LS) algorithms and offer the remarkable signal-to-residual-SI-and-noise ratio (SINR) approaching the signal-to-noise ratio (SNR).

In this chapter, we resort to a different approach for SI channel estimation and cancellation in a full-duplex transceiver in two stages. The first SI channel estimate is obtained during a short initial half-duplex period for the RF cancellation stage prior to the LNA/ADC. Noting that the SI channel has a sparse structure dominated by a relatively small number of clusters of significant paths [1], we prove that its sensing matrix satisfies the restricted isometry property (RIP) [2]. Hence, compressed-sensing (CS) theory can be applied to exploit the sparsity of the SI channel by using a mixed norm optimization criteria to return the non-zero coefficients and estimate the SI channel with much fewer samples than the linear reconstruction methods [3]. Note that CS-based channel estimation has been considered in the delay-Doppler domain, angle domain or angle delay-Doppler domain [4, 5]. We also derive the regularization parameter that can be selected to sufficiently reduce the SI.

In the second step during the full-duplex operation, the detection of the intended signal requires the knowledge of the intended channel between the two transceivers. We develop a subspace-based algorithm to jointly estimate the residual SI and

T. Le-Ngoc, A. Masmoudi, *Full-Duplex Wireless Communications Systems*, Wireless Networks, DOI 10.1007/978-3-319-57690-9_4

intended channels combined with the transmitter nonlinearity for the baseband SI-cancellation stage. Since the coefficients are obtained up to a matrix ambiguity [6], we propose a method to find the expression of the SI and intended channels ambiguity matrices with much smaller number of training samples than traditional data-aided estimator. Our algorithm is based on the orthogonality between the signal and noise subspaces, a property widely exploited in the field of array processing for parameter estimation and spectral analysis [7, 8], applied here to the problem of channel and nonlinearity estimation in full-duplex systems [9]. The reduction of training data used in the proposed estimator can be explained by the fact that the estimator exploits the information bearing in the unknown data to find the subspace of the transmit signal. The knowledge of the signal subspace reduces the number of the remaining parameters to estimate compared to the LS estimator.

There are different reasons that motivate us to develop another algorithm in the second cancellation stage different from the algorithm in the first stage. First, the residual SI channel after the first cancellation stage does not have any specific sparse structure. Second, we need to jointly estimate the residual SI and the intended channels without knowing the transmitted data from the other transceiver, and third we need to reconstruct the distorted SI signal from the estimated nonlinear coefficients of the transmitter. In this situation, the CS estimator cannot recover the channel coefficients without a perfect knowledge of the data. In previous full-duplex implementations [1, 10, 11], it is not clear how the residual SI channel is estimated in the presence of the intended signal, which acts as noise when trying to estimate the residual SI channel. The presence of the intended signal affects the baseband SI-cancellation stage, which motivates us to develop a joint estimation of the different parameters.

This chapter is organized as follows. In Sect. 4.1, the full-duplex transceiver with RF and baseband SI-cancellation stages is presented. We present our system model in Sect. 4.2. In Sect. 4.3, we develop the subspace-based technique for the joint estimation of the residual SI channel, the intended channel and take into account the transmitter impairments for the baseband SI-cancellation stage. Illustrative simulation results are given in Sect. 4.4 and Sect. 4.5 presents the conclusion.

4.1 Full-Duplex System Model

The considered MIMO-OFDM transceiver follows the general direct-conversion structure in Fig. 1.3 with N_t transmit streams and N_r receive streams operating in a full-duplex fashion. The numbers of transmit and receive streams are equal for both transceivers. Beside the Tx-Rx isolation provided in the multi-antenna sub-system,[1] two SI-cancellation stages are included. The RF SI-cancellation stage is done at

[1]The multi-antenna sub-system may include an analog RF cancellation stage to achieve large Tx-Rx isolation.

RF before LNA and ADC in order to avoid overloading/saturation. In practical implementation, the IQ mixer has some imbalance between the I and Q components, which results in an inband image of the signal. Similar to Chap. 2, the output of the transmit IQ mixer of the Tx stream q is:

$$x_q^{IQ}(t) = g_{1,q}x_q(t) + g_{2,q}x_q^*(t), \tag{4.1}$$

where $g_{1,q}$ and $g_{2,q}$ represent the responses to the direct signal and its image, respectively, and $x_q(t)$ is the transmitted signal from the Tx stream q. The complex-baseband equivalent signal $x_q^{IQ}(t)$ then passes through a nonlinear PA whose response is modeled with a Hammerstein nonlinearity as:

$$x_q^{PA}(t) = \left(\alpha_{1,q}x_q^{IQ}(t) + \sum_{p=1}^{P} \alpha_{2p+1,q}x_q^{IQ}(t)|x_q^{IQ}(t)|^{2p} \right) \star f(t), \tag{4.2}$$

where $f(t)$ models the memory of the PA, $\alpha_{1,q}$ and $\alpha_{2p+1,q}$ are the linear gain and the $(2p+1)$th-order gain, respectively, for a nonlinearity order of P and $\star$ denotes the convolution operation. In this chapter, we limit our analysis to the third-order nonlinearity ($P = 1$) to simplify the notation. Considering multipath channels, the received signal at the Rx stream r can be written as:

$$y_r^{ant}(t) = \sum_{q=1}^{N_t} h_{r,q}^i(t) \star x_q^{PA}(t) + h_{r,q}^{us}(t) \star s_q(t) + w^{(r)}(t), \tag{4.3}$$

where $s_q(t)$ is the transmitted intended signal from the Tx stream q of the other intended transmitter, extended by the cyclic prefix of length N_{cp}. $h_{r,q}^i(t)$ is the SI channel impulse response of the link from Tx stream q to Rx stream r of the *same* transceiver while $h_{r,q}^{us}(t)$ is the *intended* channel impulse response of the link from Tx stream q of the *other intended* transmitter to Rx stream r. $w^{(r)}(t)$ is the additive thermal noise in Rx stream r. We mention that these definitions are presented in Sect. 2.6, and we reproduce them here for convenience and to introduce the different notations for MIMO systems. Then, the received signal passes through an LNA whose output signal is:

$$y_r^{LNA}(t) = k_{LNA}y_r^{ant}(t) + w_{LNA}^{(r)}(t), \tag{4.4}$$

where $w_{LNA}^{(r)}(t)$ is the additive noise caused by the LNA and k_{LNA} is the gain of the LNA. Finally, the amplitude of the received signal is adjusted by a VGA to match the dynamic range of the ADC. Substituting (4.2) and (4.3) into (4.4) and assuming unity linear gain ($\alpha_{1,q} = 1$ and $g_{1,q} = 1$), the output samples are given by:

$$y_r(n) = \sum_{q=1}^{N_t}\sum_{l=0}^{L} h_{r,q}^{si}(l)x_q^{IQ}(n-l) + \alpha_{3,q}h_{r,q}^{si}(l)x_{q,ip3}(n-l) + h_{r,q}^{s}(l)s_q(n-l) + w_r(n), \tag{4.5}$$

where $(L+1)$ denotes the number of resolvable paths, the overall channel responses are defined as:

$$\begin{aligned} h_{r,q}^{si}(l) &= k_{LNA}h_{r,q}^{i}(l) \star f(l), \\ h_{r,q}^{s}(l) &= k_{LNA}h_{r,q}^{us}(l), \end{aligned} \tag{4.6}$$

$x_{q,ip3}(n) = x_q^{IQ}(n)|x_q^{IQ}(n)|^2$ and $w_r(n)$ collects the quantization noise, the LNA noise and the thermal noise. In this chapter, we suppose that the phase noise is low enough to be ignored. From (4.5), it follows that the vector $\boldsymbol{y}(n)$ can be written as:

$$\boldsymbol{y}(n) = \sum_{l=0}^{L}\left(\boldsymbol{X}_{iq}(n-l)\boldsymbol{h}^{si}(l) + \boldsymbol{X}_{ip3}(n-l)(\boldsymbol{A}_3 \otimes \boldsymbol{I}_{N_r})\boldsymbol{h}^{si}(l) + \boldsymbol{S}(n-l)\boldsymbol{h}^{s}(l)\right) + \boldsymbol{w}(n), \tag{4.7}$$

where

$$\begin{aligned} \boldsymbol{y}(n) &= [y_1(n),\ y_2(n), \ldots,\ y_{N_r}(n)]^T, \\ \boldsymbol{h}^{si}(l) &= \left[\boldsymbol{h}_1^{siT}(l), \ldots,\ \boldsymbol{h}_{N_t}^{siT}(l)\right]^T, \\ \boldsymbol{h}_q^{si}(l) &= \left[h_{1,q}^{si}(l),\ h_{2,q}^{si}(l), \ldots,\ h_{N_r,q}^{si}(l)\right]^T, \\ \boldsymbol{h}^{s}(l) &= \left[\boldsymbol{h}_1^{sT}(l), \ldots,\ \boldsymbol{h}_{N_t}^{sT}(l)\right]^T, \\ \boldsymbol{h}_q^{s}(l) &= [h_{1,q}^{s}(l),\ h_{2,q}^{s}(l), \ldots,\ h_{N_r,q}^{s}(l)]^T, \\ \boldsymbol{A}_3 &= \mathrm{diag}\{\alpha_{3,1},\ \alpha_{3,2}, \ldots,\ \alpha_{3,N_t}\}, \\ \boldsymbol{w}(n) &= [w_1(n),\ w_2(n), \ldots,\ w_{N_r}(n)]^T. \end{aligned} \tag{4.8}$$

In (4.7), $\boldsymbol{X}_{iq}(n)$ is a $N_r \times N_t N_r$ Toeplitz matrix with the first column given by the $N_r \times 1$ vector $\left[x_1^{IQ}(n),\ 0, \ldots,\ 0\right]$ and the first row given by $\left[x_1^{IQ}(n),\ x_2^{IQ}(n), \ldots,\ x_{N_t}^{IQ}(n)\right] \otimes \boldsymbol{e}_1$ with $\boldsymbol{e}_1$ being the $1 \times N_r$ vector having one in the first element and zeroes elsewhere. The matrices $\boldsymbol{X}_{ip3}(n)$ and $\boldsymbol{S}(n)$ are constructed in the same way as $\boldsymbol{X}_{iq}(n)$ but with the samples $x_{q,ip3}(n)$ and $s_q(n)$, respectively.

Now let the two $N_t N_r(L+1) \times 1$ vectors $\boldsymbol{h}^{si}$ and $\boldsymbol{h}^{s}$ gather all the coefficients of the SI and intended channels, respectively, i.e.,

$$\begin{aligned} \boldsymbol{h}^{si} &= \left[\boldsymbol{h}^{siT}(0),\ \boldsymbol{h}^{siT}(1), \ldots,\ \boldsymbol{h}^{siT}(L)\right]^T, \\ \boldsymbol{h}^{s} &= \left[\boldsymbol{h}^{sT}(0),\ \boldsymbol{h}^{sT}(1), \ldots,\ \boldsymbol{h}^{sT}(L)\right]^T, \end{aligned} \tag{4.9}$$

and define:

$$
X_{iq} = \begin{pmatrix}
X_{iq}(0) & X_{iq}(N-1) \dots & X_{iq}(N-L) \\
X_{iq}(1) & X_{iq}(0) \quad \ddots & \vdots \\
\vdots & \ddots & X_{iq}(N-1) \\
\vdots & & X_{iq}(0) \\
\vdots & & \vdots \\
X_{iq}(N-1) & X_{iq}(N-2) \dots & X_{iq}(N-L-1)
\end{pmatrix},
$$

$$
S = \begin{pmatrix}
S(0) & S(N-1) \dots & S(N-L) \\
S(1) & S(0) \quad \ddots & \vdots \\
\vdots & \ddots & S(N-1) \\
\vdots & & S(0) \\
\vdots & & \vdots \\
S(N-1) & S(N-2) \dots & S(N-L-1)
\end{pmatrix}. \tag{4.10}
$$

The $N_r N \times N_t N_r (L+1)$ *self* signal matrix X_{iq} includes samples transmitted of the OFDM symbol from the same transceiver affected by the IQ mixer and the $N_r N \times N_t N_r (L+1)$ *intended* signal matrix S contains samples transmitted from the other intended transmitter. Then, the received $N_r N \times 1$ vector $y = \left[y^T(0), \dots, y^T(N-1)\right]^T$, after removing the cyclic prefix, is given by:

$$
y = X_{iq} h^{si} + X_{ip3}(I_{L+1} \otimes A_3 \otimes I_{N_r}) h^{si} + S h^{s} + w, \tag{4.11}
$$

where X_{ip3} is defined in the same way as X_{iq} in (4.10) and w is the $N_r N \times 1$ thermal noise vector.

In full-duplex systems, the SI, shown by the first and the second terms in (4.11), is many orders of magnitude higher than the intended signal from the other intended transmitter, shown by the third term in (4.11). This imposes different cancellation stages to reduce the SI to a sufficiently low level for proper signal detection [1]. The RF cancellation stage aims to suppress the SI prior to the receiver's LNA/ADC. Since the transmitted signal is known, we only need to estimate the SI channel h^{si} to generate the SI replica at RF for cancellation. The remaining SI and the nonlinear terms after the receiver's ADC will be further suppressed by the baseband cancellation stage as shown in Fig. 3.1. The proposed estimation and cancellation algorithms for the RF and baseband cancellation stages will be discussed in Sects. 4.2 and 4.3, respectively.

4.2 Compressed-Sensing-Based RF Cancellation Stage

As previously discussed, one major task in the RF cancellation stage is to estimate the SI channel vector $\boldsymbol{h}^{si}$. In this section, we build the matrix $\boldsymbol{X}$, having the same form as $\boldsymbol{X}_{iq}$ in (4.10) from the reference SI signal. If we suppose that an estimate of the SI channel is available, say at time index t, the SI replica is generated and subtracted from the received signal in (4.11) during the next transmitted OFDM symbol at time index $t+1$ to obtain:

$$\widetilde{\boldsymbol{y}}_{t+1} = \boldsymbol{y}_{t+1} - \boldsymbol{X}_{t+1}\widehat{\boldsymbol{h}}^{si}. \tag{4.12}$$

In order to suppress the SI at time index $t+1$, an estimate of the SI channel should be available from earlier. Therefore, a half-duplex transmission period[2] is needed at the beginning to estimate the SI channel and then to reduce the SI without affecting the intended signal when switching to full-duplex transmission.

During the initial half-duplex fashion period, the transceiver receives only its own signal. The signal model in (4.11) reduces to:

$$\boldsymbol{y} = \boldsymbol{X}\boldsymbol{h}^{si} + \boldsymbol{w}, \tag{4.13}$$

where the time index is omitted for clarity. The estimation of the SI channel $\boldsymbol{h}^{si}$ is equivalent to the traditional problem of training based channel estimation. Usually, the algorithms to solve this problem rely on the linear LS strategies [12–14]. However, these methods do not exploit the particular structure of the channel. Actually, the SI channel exhibits a sparse structure coming from the fact that the largest tap delay is usually much larger than the number of nonzero taps. For example, consider the simple and popular architecture using the same antenna to transmit and receive via a 3-port circulator, the dominant paths of the SI channel come from the leakage through the circulator and the internal antenna reflection due to the impedance-mismatch between isolator and antenna. On the other hand, external reflection from closely located objects may occur with much larger delay and weaker level as compared to the two dominant paths since they travel longer distance [15]. The "zeros" are actually located between the reflections and the dominant paths and we do not have any a priori information about the delays of the reflected paths. When using two different antennas to transmit and receive, the LoS components and the path coming from the electromagnetic waves reflected from the transceiver structure have delays shorter than 3 ns [16, 17]. Therefore, the channel impulse response has multiple peaks with relatively higher amplitude for the LoS path and a number of zeros between two consecutive paths representing the delay difference between the two paths. This behaviour has been verified by channel measurements performed by our research group. Therefore, the problem turns out

---

[2] While this initial period is used as a training period to estimate $\boldsymbol{h}^{si}$, two-way communications are in a half-duplex fashion.

to estimating a sparse channel from the observation $\boldsymbol{y}$. Hence, mathematically, we are looking for $\arg\min_{\boldsymbol{h}} ||\boldsymbol{h}||_0$ such that $\boldsymbol{y} = \boldsymbol{X}\boldsymbol{h}$. This is, however, a difficult combinatorial optimization problem and may be intractable even for small size problem. Recently, it has been shown that when $\boldsymbol{h}$ is sparse, it is possible to replace $||\boldsymbol{h}||_0$ by $||\boldsymbol{h}||_1$ in the optimization problem and we still obtain the exact same solutions for both problems [18]. The new problem:

$$\arg\min_{\boldsymbol{h}} ||\boldsymbol{h}||_1 \text{ such that } \boldsymbol{y} = \boldsymbol{X}\boldsymbol{h}, \tag{4.14}$$

is a convex optimization problem and can be solved by linear programming. In practice, only noisy measurements are available. Therefore, the constraint $\boldsymbol{y} = \boldsymbol{X}\boldsymbol{h}$ is replaced by $||\boldsymbol{y} - \boldsymbol{X}\boldsymbol{h}||_2^2 \leq \lambda$, for some parameter λ, to introduce the additive noise. This optimization problem is computationally tractable since it can be recast as a second-order cone programming [3].

To include the transmitter nonlinearity when cancelling the SI in the RF, the output of the transmitter chain is taken as a reference signal and convolved with the estimated channel [9, 10, 19, 20]. That is, if we are able to obtain the exact value of $\boldsymbol{h}^{si}$, we will have $\left|\left|\boldsymbol{y} - \boldsymbol{X}_{iq}\boldsymbol{h}^{si} - \boldsymbol{X}_{ip3}(\boldsymbol{I}_{L+1} \otimes \boldsymbol{A}_3 \otimes \boldsymbol{I}_{N_r})\boldsymbol{h}^{si}\right|\right|_2^2 = \left|\left|\boldsymbol{w}\right|\right|_2^2$ which can be approximated by $\sigma^2 N_r N$ for sufficiently large noise vector $\boldsymbol{w}$, where σ^2 is the noise variance. However, the estimated value $\widehat{\boldsymbol{h}}$ cannot exactly match the real channel $\boldsymbol{h}^{si}$. Let $\boldsymbol{h}^{rsi}$ denote the residual channel ($\boldsymbol{h}^{rsi} = \boldsymbol{h}^{si} - \widehat{\boldsymbol{h}}$). In that case, we have:

$$\boldsymbol{y} - \boldsymbol{X}_{iq}\widehat{\boldsymbol{h}} - \boldsymbol{X}_{ip3}(\boldsymbol{I}_{L+1} \otimes \boldsymbol{A}_3 \otimes \boldsymbol{I}_{N_r})\widehat{\boldsymbol{h}} = \boldsymbol{X}_{iq}\boldsymbol{h}^{rsi} + \boldsymbol{X}_{ip3}(\boldsymbol{I}_{L+1} \otimes \boldsymbol{A}_3 \otimes \boldsymbol{I}_{N_r})\boldsymbol{h}^{rsi} + \boldsymbol{w}, \tag{4.15}$$

After the RF SI-cancellation stage, the residual SI should be sufficiently low to avoid saturating the receiver LNA/ADC and to allow the successful signal detection in the baseband. Therefore, using the estimated vector $\widehat{\boldsymbol{h}^{si}}$, we want to obtain:

$$\begin{aligned}\left|\left|\boldsymbol{y} - \boldsymbol{X}_{iq}\widehat{\boldsymbol{h}} - \boldsymbol{X}_{ip3}(\boldsymbol{I}_{L+1} \otimes \boldsymbol{A}_3 \otimes \boldsymbol{I}_{N_r})\widehat{\boldsymbol{h}}\right|\right|_2^2 &= \left|\left|\boldsymbol{X}_{iq}\boldsymbol{h}^{rsi} + \boldsymbol{X}_{ip3}(\boldsymbol{I}_{L+1} \otimes \boldsymbol{A}_3 \otimes \boldsymbol{I}_{N_r})\boldsymbol{h}^{rsi} + \boldsymbol{w}\right|\right|_2^2 \\ &\approx (P_s + \sigma^2)N_r N,\end{aligned} \tag{4.16}$$

where P_s is the power of the received intended signal. To that end, the regularization parameter λ is chosen to be high enough so that $(P_s + \sigma^2)N_r N \approx \lambda$.

The attractive feature in CS theory is that a smaller number of measurements than the length of $\boldsymbol{h}^{si}$ is sufficient to recover $\boldsymbol{h}^{si}$. This reconstruction ability depends on the matrix $\boldsymbol{X}$. In particular, it suffices that the matrix $\boldsymbol{X}$ satisfies the restricted isometry property (RIP) introduced in [2] as follows. Let S denotes the number of non-zero elements in the vector $\boldsymbol{h}^{si}$. According to the definition in [2], $\boldsymbol{X}$ satisfies the RIP[3] of order $2S$ with parameter $\delta_S \in [0,\ 1]$, for a given integer S, if for every vector $\boldsymbol{\theta}$ such that $||\boldsymbol{\theta}||_0 \leq 2S$ we have:

[3]The RIP guarantees the uniqueness of the solution to the problem. In fact, for any two different S sparse vectors $\boldsymbol{\theta}_1$ and $\boldsymbol{\theta}_2$, the vector $\boldsymbol{\theta}_1 - \boldsymbol{\theta}_2$ has at most $2S$ non zeros elements (if the non-zero

$$(1-\delta_S)||\boldsymbol{\theta}||_2^2 \leq ||\boldsymbol{X}\boldsymbol{\theta}||_2^2 \leq (1+\delta_S)||\boldsymbol{\theta}||_2^2. \tag{4.17}$$

In other words, $\boldsymbol{X}$ satisfies the RIP if the singular values of all the submatrices $\boldsymbol{X}_{\mathcal{T}}$, formed from $\boldsymbol{X}$ by taking the columns indexed by $\mathcal{T}$ from $\boldsymbol{X}$, are in $\left[\sqrt{1-\delta_S},\ \sqrt{1+\delta_S}\right]$, where $\mathcal{T} \subset \{1,\ldots,N_tN_r(L+1)\}$ with cardinality no larger than S. It follows that, to prove the RIP for a given matrix, it suffices to bound the eigenvalues of the $S\times S$ Gramian matrix $\boldsymbol{G}_{\mathcal{T}} = \boldsymbol{X}_{\mathcal{T}}^H\boldsymbol{X}_{\mathcal{T}}$ in the interval $[1-\delta_S,\ 1+\delta_S]$, for all subsets of column indices $\mathcal{T}$. According to the Geršgorin's Disc theorem [21], the eigenvalues of $\boldsymbol{G}_{\mathcal{T}}$ lie in the union of the S discs d_i centered at $c_i = \boldsymbol{G}_{\mathcal{T}}(i,i)$ and with radius $r_i = \sum_{j\neq i,j=1}^{S} |\boldsymbol{G}_{\mathcal{T}}(i,j)|$, for $i = 1,\ldots,S$. That is, for two δ_d and δ_o real in $[0,\ 1]$ and satisfying $\delta_d + \delta_o = \delta_S$, if all the diagonal elements of $\boldsymbol{G}_{\mathcal{T}}$ verify $|\boldsymbol{G}_{\mathcal{T}}(i,i) - 1| < \delta_d$ and all the off-diagonal elements satisfy $|G_{\mathcal{T}}(i,j)| < \delta_o/S$, then all the eigenvalues of $\boldsymbol{G}_{\mathcal{T}}$ contained in the union of the discs d_i, $i = 1,\ldots,\ S$, are in the range $[1-\delta_S,\ 1+\delta_S]$.

Proof of the RIP: We need to establish bounds on $|\boldsymbol{G}_{\mathcal{T}}(i,i) - 1|$ and $\sum_{j=1,j\neq i}^{S} |\boldsymbol{G}_{\mathcal{T}}(i,j)|$, for all subsets $\mathcal{T}$. In the following proof, the elements of $\boldsymbol{X}$ are Gaussian random variables with mean 0 and variance $1/N$. The matrix $\boldsymbol{X}$ also verifies the RIP when its elements have arbitrary variance σ_x^2 by multiplying each term in the inequality (4.17) by N/σ_x^2. Moreover, we suppose a real matrix $\boldsymbol{X}$.

Using Lemma 5 in [21], each diagonal element of $\boldsymbol{G}_{\mathcal{T}}(i,i) = \sum_{n=0}^{N-1} |x_{p_i}(n)|^2$, where p_i is the corresponding transmit antenna index, verifies:

$$\Pr\left(|\boldsymbol{G}_{\mathcal{T}}(i,i) - 1| \geq \delta_d\right) \leq 2\exp\left(-\frac{N\delta_d}{16}\right), \tag{4.18}$$

where $\Pr(A)$ is the probability of the event A. Each column of $\boldsymbol{X}$ contains the N transmitted samples from one of the N_t transmitted streams. Therefore, there are exactly N_t different values for $\boldsymbol{G}_{\mathcal{T}}(i,i)$. By the union bound, we have for every subset $\mathcal{T}$ and for all $i = 1,\ldots,\ S$:

$$\Pr\left(\bigcup_{\mathcal{T}}\bigcup_{i=1}^{S} |\boldsymbol{G}_{\mathcal{T}}(i,i) - 1| \geq \delta_d\right) \leq 2N_t\exp\left(-\frac{N\delta_d}{16}\right). \tag{4.19}$$

For a given subset $\mathcal{T}$, any off-diagonal element $\boldsymbol{G}_{\mathcal{T}}(i,j)$ is the inner product between the m_i and m_j columns of $\boldsymbol{X}$. For convenience, we write $m_i = n_i + p_iN_r + d_iN_rN_t$ with $n_i \in [1,\ N_r]$, $p_i \in [0,\ N_t - 1]$ and $d_i \in [0,\ L]$. Depending on m_i and m_j, we distinguish the following different cases:

1. If $n_i \neq n_j$, then $\boldsymbol{G}_{\mathcal{T}}(i,j) = 0$.
2. If $n_i = n_j$ and $d_i = d_j$ then $\boldsymbol{G}_{\mathcal{T}}(i,j)$ is the sum of N terms given by $\boldsymbol{G}_{\mathcal{T}}(i,j) = \sum_{n=0}^{N-1} x_{p_i+1}(n)x_{p_j+1}(n)$. The entries of the previous summation are independent.

elements of $\boldsymbol{\theta}_1$ and $\boldsymbol{\theta}_2$ are not in the same positions). According to the RIP inequality, the two images of $\boldsymbol{\theta}_1$ and $\boldsymbol{\theta}_2$ are different as long as $\boldsymbol{\theta}_1$ is different from $\boldsymbol{\theta}_2$.

Therefore, applying Lemma 6 in [21] we obtain the following bound:

$$\Pr\left(|\boldsymbol{G}_{\mathcal{T}}(i,j)| \geq \frac{\delta_S}{S}\right) \leq 2\exp\left(-\frac{\delta_o^2 N}{4S^2\left(1+\frac{\delta_o}{2S}\right)}\right). \tag{4.20}$$

The total number of unique elements having this form is $\frac{N_t^2-N_t}{2}$.

3. If $n_i = n_j$, $d_i \neq d_j$ and $p_i \neq p_j$, then $\boldsymbol{G}_{\mathcal{T}}(i,j) = \sum_{n=0}^{N-1-|d_i-d_j|} x_{p_i+1}(n)x_{p_j+1}(n + |d_i - d_j|)$ is the sum of $N - |d_i - d_j|$ independent terms. Using the same formula as in case 2 gives:

$$\Pr\left(|\boldsymbol{G}_{\mathcal{T}}(i,j)| \geq \frac{\delta_S}{S}\right) \leq 2\exp\left(-\frac{\delta_o^2 N}{4S^2\left(\frac{N-|d_i-d_j|}{N}+\frac{\delta_o}{2S}\right)}\right). \tag{4.21}$$

There is $L(N_t^2 - N_t)/2$ different terms having this form.

4. If $n_i = n_j$, $d_i \neq d_j$ and $p_i = p_j$, then $\boldsymbol{G}_{\mathcal{T}}(i,j)$ is given by:

$$\boldsymbol{G}_{\mathcal{T}}(i,j) = \sum_{n=0}^{N-1-|d_i-d_j|} x_{p_i+1}(n)x_{p_i+1}(n + |d_i - d_j|). \tag{4.22}$$

Unlike the other cases, the entries of the summation are no longer independent since each element $x_{p_i+1}(n)$ appears in two entries. For example, consider that $|d_i - d_j| = 1$, then we have:

$$\begin{aligned}\boldsymbol{G}_{\mathcal{T}}(i,j) = x_{p_i+1}(1)x_{p_i+1}(0) + x_{p_i+1}(2)x_{p_i+1}(1) + x_{p_i+1}(3)x_{p_i+1}(2)\ldots\\ + x_{p_i+1}(N-1)x_{p_i+1}(N-2).\end{aligned} \tag{4.23}$$

Since the odd-order terms are mutually independent, and the even-order terms are also mutually independent, the summation in (4.23) can be split into two sums, each for the mutually independent variables. Therefore:

$$\begin{aligned}\Pr\left(|\boldsymbol{G}_{\mathcal{T}}(i,j)| \geq \frac{\delta_o}{S}\right) &\leq \Pr\left(|\boldsymbol{G}^1_{\mathcal{T}}(i,j)| \geq \frac{\delta_o}{2S} \text{ or } |\boldsymbol{G}^2_{\mathcal{T}}(i,j)| \geq \frac{\delta_o}{2S}\right)\\ &\leq 2\max\left(\Pr\left(|\boldsymbol{G}^1_{\mathcal{T}}(i,j)| \geq \frac{\delta_o}{2S}\right), |\boldsymbol{G}^2_{\mathcal{T}}(i,j)| \geq \frac{\delta_o}{2S}\right)\\ &\leq 4\exp\left(-\frac{\delta_o^2 N}{6S^2}\right),\end{aligned} \tag{4.24}$$

where the last equality follows from the upper bound used in (4.21).

We gather the previous results along with the union bound to establish an upper bound on the probability that all the elements $\boldsymbol{G}_{\mathscr{T}}(i,j)$, for any subset $\mathscr{T}$ and $i \neq j$, satisfy $|\boldsymbol{G}_{\mathscr{T}}(i,j)| \geq \frac{\delta_o}{S}$:

$$\Pr\left(\bigcup_{\mathscr{T}}\bigcup_{j=1}^{S}|\boldsymbol{G}_{\mathscr{T}}(i,j)|\geq\frac{\delta_o}{S}\right)\leq 2(L+1)N_t^2\exp\left(-\frac{\delta_o^2 N}{6S^2}\right). \tag{4.25}$$

To obtain the result claimed in Sect. 4.2, let $\delta_d = 2\delta_S/3$, $\delta_o = \delta_S/3$ and use (4.19) and (4.25) to obtain:

$$\begin{aligned}\Pr(\boldsymbol{X}\text{ not satisfying RIP}) &\leq 2(L+1)N_t^2\exp\left(-\frac{\delta_S^2 N}{54S^2}\right)+2N_t\exp\left(-\frac{N\delta_S}{36}\right)\\ &\leq \left(2(L+1)N_t^2+2N_t\right)\exp\left(-\frac{\delta_S^2 N}{54S^2}\right).\end{aligned} \tag{4.26}$$

Define $c_1 = 2(L+1)N_t^2 + 2N_t$ and for $c_2 < \delta_S^2/54$, we obtain:

$$\Pr(\boldsymbol{X}\text{ not satisfying RIP}) \leq \exp\left(-\frac{c_2 N}{S^2}\right), \tag{4.27}$$

for any $N \geq \frac{54S^2\log(c_1)}{-54c_2+\delta_S^2}$. It follows that the matrix $\boldsymbol{X}$ satisfies the RIP with parameter δ_S with probability exceeding:

$$1-\exp\left(-\frac{c_2 N}{S^2}\right). \tag{4.28}$$

4.3 Subspace-Based Baseband Cancellation Stage

Once the two-way communications start full-duplex operation, the SI channel estimate obtained during the training period is used to reduce the power of the SI. After the RF cancellation stage, the resulting signal in baseband is given by:

$$\boldsymbol{y}_c(n)=\sum_{q=1}^{N_t}\sum_{l=0}^{L}\left(\boldsymbol{h}_q^{rsi}(l)x_q^{IQ}(n-l)+\alpha_{3,q}\boldsymbol{h}_q^{rsi}(l)x_{q,ip3}(n-l)+\boldsymbol{h}_q^{s}(l)s_q(n-l)\right)+\boldsymbol{w}(n), \tag{4.29}$$

where we use the similar vector structures in Sect. 4.1. For the RF cancellation stage, the reference signal is taken after the transmit PA [10, 20, 22]. Therefore, the transmitter impairments are included in the reference signal and consequently, only the SI channel is needed to model the received SI. On the other hand, the reference signal for the baseband cancellation is taken from the modulator and thus does not contain the transmitter impairments. As a consequence, we need to estimate the residual SI channel as well as the transmitter impairments. Since the self-signal is known, the simplest way to estimate the corresponding coefficients is to resort to a linear estimator. But this method will suffer from large estimation error since the intended signal appears as additive noise and cannot suppress the resulting distortions from the transmitter impairments. Therefore, the intended signal and the nonlinear SI components should also be considered in the estimation process. In this section, we develop a subspace-based method for jointly estimating the SI and intended channels and the nonlinearity coefficients.

Before presenting the proposed estimator, we need to have a more tractable representation of the received signal $\boldsymbol{y}_c(n)$ to introduce our algorithm. By defining:

$$\begin{aligned}
\boldsymbol{x}(n) &= \left[x_1^{IQ}(n) + \alpha_{3,1}x_{1,ip3}(n), \ldots,\ x_{N_t}^{IQ}(n) + \alpha_{3,N_t}x_{N_t,ip3}(n)\right]^T, \\
\boldsymbol{s}(n) &= \left[s_1(n),\ s_2(n), \ldots,\ s_{N_t}(n)\right]^T, \\
\boldsymbol{H}^{rsi}(l) &= \left[\boldsymbol{h}_1^{rsi}(l),\ \boldsymbol{h}_2^{rsi}(l), \ldots,\ \boldsymbol{h}_{N_t}^{rsi}(l)\right], \\
\boldsymbol{H}^{s}(l) &= \left[\boldsymbol{h}_1^{s}(l),\ \boldsymbol{h}_2^{s}(l), \ldots,\ \boldsymbol{h}_{N_t}^{s}(l)\right],
\end{aligned} \tag{4.30}$$

the cancelled input signal $\boldsymbol{y}_c(n)$ can be expressed as:

$$\boldsymbol{y}_c(n) = \sum_{l=0}^{L} \boldsymbol{H}^{rsi}(l)\boldsymbol{x}(n-l) + \boldsymbol{H}^{s}(l)\boldsymbol{s}(n-l) + \boldsymbol{w}(n). \tag{4.31}$$

Then, we gather the two channel matrices $\boldsymbol{H}^{s}(l)$ and $\boldsymbol{H}^{rsi}(l)$ in one matrix $\boldsymbol{H}(l) = \left[\boldsymbol{H}^{rsi}(l)\ \boldsymbol{H}^{s}(l)\right]$ and define the $N_rM \times 2N_tN$ block Toeplitz matrix:

$$\boldsymbol{H} = \begin{pmatrix}
\boldsymbol{0} & \ldots\ \boldsymbol{0} & \boldsymbol{H}(L) & \ldots & \boldsymbol{H}(0) & & \\
\vdots & & & \ddots & & \ddots & \\
\boldsymbol{0} & & & & \boldsymbol{0}\ \boldsymbol{H}(L) & \ldots & \boldsymbol{H}(0) \\
\boldsymbol{H}(0) & & & & & & \boldsymbol{H}(1) \\
\vdots & & & & & \ddots & \\
& \ddots & & & & & \boldsymbol{H}(L) \\
\boldsymbol{H}(L) & & \ddots & & & & \boldsymbol{0} \\
\boldsymbol{0} & & & & & & \\
& \ddots & & \boldsymbol{H}(0) & & & \vdots \\
& & \ddots & & & \ddots & \\
& & & \vdots & & & \boldsymbol{0} \\
\boldsymbol{0} & & & & \boldsymbol{0}\ \boldsymbol{H}(L) & \ldots & \boldsymbol{H}(0)
\end{pmatrix}, \tag{4.32}$$

where $M = N + N_{cp} - L$ and the transmitted data in one $2N_tN \times 1$ vector:

$$\boldsymbol{u} = \left[\boldsymbol{x}^T(0),\ \boldsymbol{s}^T(0), \ldots,\ \boldsymbol{x}^T(N-1),\ \boldsymbol{s}^T(N-1)\right]^T. \tag{4.33}$$

Using these notations, the received N_rM vector over the N_r antennas is given by:

$$\begin{aligned}\boldsymbol{y}_c &= \left[\boldsymbol{y}_c^T(-N_{cp}+L), \ldots,\ \boldsymbol{y}_c^T(-1),\ \boldsymbol{y}_c^T(0),\ \boldsymbol{y}_c^T(1), \ldots,\ \boldsymbol{y}_c^T(N-1)\right]^T \\ &= \boldsymbol{Hu} + \boldsymbol{w},\end{aligned} \tag{4.34}$$

where the negative index refers to the cyclic prefix part of the received signal. Note that for multi-block transmission, the vector in (4.34) is indexed according to the block number t, i.e., $\boldsymbol{y}_{c,t}$. We omit this indexation for simplicity and we consider a given number of block to later estimate the covariance matrix of $\boldsymbol{y}_c$.

4.3.1 Subspace-Based Technique

We assume that the noise samples are uncorrelated, i.e., $\mathbb{E}\{w(n)w^*(m)\} = \sigma^2$ if $n = m$ and 0 if $n \neq m$, and the noise and signal samples are also uncorrelated. It follows that, the covariance matrix $\boldsymbol{R}_{y_c}$ of $\boldsymbol{y}_c$ is given by:

$$\begin{aligned}\boldsymbol{R}_{y_c} &= \mathbb{E}\left\{\boldsymbol{y}_c\boldsymbol{y}_c^H\right\} \\ &= \boldsymbol{HR}_u\boldsymbol{H}^H + \sigma^2\boldsymbol{I}_{N_rM},\end{aligned} \tag{4.35}$$

where $\boldsymbol{R}_u$ is the $2NN_t \times 2NN_t$ covariance matrix of $\boldsymbol{u}$.

In practice, the sample estimate, $\widehat{\boldsymbol{R}}_{y_c}$ of the covariance matrix $\boldsymbol{R}_{y_c}$ is used in the estimation process. Considering T transmit OFDM symbols, $\widehat{\boldsymbol{R}}_{y_c}$ is obtained by a time-average:

$$\widehat{\boldsymbol{R}}_{y_c} = \frac{1}{T}\sum_{t=1}^{T}\boldsymbol{y}_{c,t}\boldsymbol{y}_{c,t}^H. \tag{4.36}$$

The signal subspace is the span of the columns of the matrix $\boldsymbol{H}$ and the noise subspace is the orthogonal complement to the signal subspace. By assuming independent channels between different antennas, the dimension of the signal subspace is $2NN_t$ (i.e., the rank of $\boldsymbol{HR}_u\boldsymbol{H}^H$ is $2NN_t$) and the dimension of the noise subspace is $p = N_rM - 2NN_t$ [23]. To guarantee that the noise subspace is nondegenerate ($p > 0$), the number of transmitting antenna in each transceiver N_t should be smaller than[4] $\lfloor \frac{N_rM}{2N} \rfloor$. Therefore, the matrix $\boldsymbol{R}_{y_c}$ has p co-orthogonal

[4] $\lfloor x \rfloor$ rounds the real x to the nearest integer smaller or equal to x.

eigenvectors, denoted by $\boldsymbol{v}_i,\ i = 1, 2, \ldots, p$ corresponding to the smallest eigenvalue of $\boldsymbol{R}_{y_c}$, i.e., σ^2. A method to avoid the additional constraint on the number of transmit and receive antennas is detailed in Chap. 5.

As the signal subspace is spanned by the $2NN_t$ columns of the matrix $\boldsymbol{H}$ and by orthogonality between the signal and noise subspaces, the columns of $\boldsymbol{H}$ are orthogonal to any vector in the noise subspace. Then we have:

$$\boldsymbol{v}_i^H \boldsymbol{H} = \boldsymbol{0}, \text{ for } i = 1, 2, \ldots, p. \tag{4.37}$$

From (4.37), we conclude that $\boldsymbol{v}_i$ spans the left null space of $\boldsymbol{H}$. Knowing the left null space of $\boldsymbol{H}$, it is possible to determine the space spanned by the columns of $\boldsymbol{H}$, denoted by span($\boldsymbol{H}$), i.e., the space containing all the linear combinations of the columns of $\boldsymbol{H}$. Therefore, knowing the span($\boldsymbol{H}$) does not give the exact matrix $\boldsymbol{H}$ since there are infinitely many matrices satisfying (4.37). However, for the specific block Toeplitz matrix that we have at hand in (4.32), it can be shown that if two matrices $\boldsymbol{H}_1$ and $\boldsymbol{H}_2$ have the same form as in (4.32) and satisfy the conditions in (4.37), then there exists a non-singular $2N_t \times 2N_t$ matrix $\boldsymbol{C}$ satisfying:

$$\boldsymbol{H}_1 = \boldsymbol{H}_2 \begin{pmatrix} \boldsymbol{C} & & & \\ & \boldsymbol{C} & & \\ & & \ddots & \\ & & & \boldsymbol{C} \end{pmatrix}. \tag{4.38}$$

The proof of the existence of $\boldsymbol{C}$ is similar to that presented in [23] with the additional condition of $\boldsymbol{H}(0)$ being full rank matrix.[5]

Recall that we are looking for a matrix that satisfies the set of equations in (4.37). Since the matrix $\boldsymbol{H}$ is entirely defined by the matrices $\boldsymbol{H}(0), \ldots, \boldsymbol{H}(L)$, instead of looking for the whole $N_rM \times 2N_tN$ matrix $\boldsymbol{H}$, we can restrict our search for the $N_r \times 2N_t$ matrices $\boldsymbol{H}(l),\ l = 0, \ldots, L$. Now, considering again the set of equations in (4.37), each eigenvector $\boldsymbol{v}_i$ can be written as:

$$\boldsymbol{v}_i = \left[\boldsymbol{v}_i^T(1),\ \boldsymbol{v}_i^T(2), \ldots,\ \boldsymbol{v}_i^T(M)\right]^T, \tag{4.39}$$

where $\boldsymbol{v}_i(m)$, for $m = 1, \ldots, M$, are $N_r \times 1$ vectors. Then, each equation in (4.37) is rearranged as:

$$\sum_{l=0}^{L} \boldsymbol{v}_i^H(n+l)\boldsymbol{H}(l) = \boldsymbol{0}, \qquad n = N_{cp} - L + 1, \ldots, N - L$$

[5]In [23], the authors proved that two Toeplitz matrices spanning the same subspace and having all zero elements above the principal diagonal are proportional with a scalar constant of proportionality. In our case, it turns out that the two matrices are related by a block diagonal matrix.

$$\sum_{l=0}^{\min(L,M-n)} \boldsymbol{v}_i^H(n+l)\boldsymbol{H}(l) + \sum_{l=\max(0,N-n+1)}^{L} \boldsymbol{v}_i^H(n-N+l)\boldsymbol{H}(l) = \boldsymbol{0}, n = N+1-L, \ldots, M, \tag{4.40}$$

or in the following matrix form:

$$\boldsymbol{\Theta}_i \check{\boldsymbol{H}} = \boldsymbol{0},\ i = 1, \ldots, p, \tag{4.41}$$

where $\check{\boldsymbol{H}} = \left[\boldsymbol{H}^T(0),\ \boldsymbol{H}^T(1), \ldots,\ \boldsymbol{H}^T(L)\right]^T$, and:

$$\boldsymbol{\Theta}_i = \begin{pmatrix} \boldsymbol{v}_i^H(N_{cp}-L+1) & \boldsymbol{v}_i^H(N_{cp}-L+2) & \ldots & \boldsymbol{v}_i^H(N_{cp}+1) \\ \boldsymbol{v}_i^H(N_{cp}-L+2) & \boldsymbol{v}_i^H(N_{cp}-L+3) & \ldots & \boldsymbol{v}_i^H(N_{cp}+2) \\ \vdots & \vdots & \ddots & \vdots \\ \boldsymbol{v}_i^H(M-L) & \boldsymbol{v}_i^H(M-L+1) & \ldots & \boldsymbol{v}_i^H(M) \\ \boldsymbol{v}_i^H(M-L+1) & \ldots & \boldsymbol{v}_i^H(M) & \boldsymbol{0} \\ \vdots & \iddots & & \vdots \\ \boldsymbol{v}_i^H(M) & \boldsymbol{0} & & \boldsymbol{0} \end{pmatrix} + \begin{pmatrix} \boldsymbol{0} & & \ldots & \boldsymbol{0} \\ \vdots & & & \vdots \\ & & \ldots & \boldsymbol{0} \\ & & & \boldsymbol{v}_i^H(1) \\ & & & \vdots \\ \boldsymbol{0} & \iddots & & \\ \boldsymbol{v}_i^H(1) & \boldsymbol{v}_i^H(2) & \ldots\ldots & \boldsymbol{v}_i^H(L+1) \\ \boldsymbol{v}_i^H(2) & \boldsymbol{v}_i^H(3) & & \boldsymbol{v}_i^H(L+2) \\ \vdots & & & \vdots \\ \boldsymbol{v}_i^H(N_{cp}-L) & \boldsymbol{v}_i^H(N_{cp}-L+1) & \ldots & \boldsymbol{v}_i^H(N_{cp}) \end{pmatrix}.$$

Collecting all the $\boldsymbol{\Theta}_i$ matrices in a $Np \times N_r(L+1)$ matrix:

$$\boldsymbol{\Theta} = \left[\boldsymbol{\Theta}_1^T,\ \boldsymbol{\Theta}_2^T, \ldots,\ \boldsymbol{\Theta}_p^T\right]^T, \tag{4.42}$$

we can rewrite (4.41) in a more compact form as:

$$\boldsymbol{\Theta}\check{\boldsymbol{H}} = \boldsymbol{0}. \tag{4.43}$$

Therefore, the column of $\check{\boldsymbol{H}}$ can be obtained by finding a basis of the null space of $\boldsymbol{\Theta}$, with the additional condition of $\check{\boldsymbol{H}} \neq \boldsymbol{0}$ to avoid the all zeroes solution. In practice, we perform the singular value decomposition (SVD) of $\boldsymbol{\Theta}$ and choose the $2N_t$ right singular vectors as the columns of $\check{\boldsymbol{H}}$.

As discussed above, the solution is not unique. For $\check{\boldsymbol{H}}_0$ obtained from the SVD of $\boldsymbol{\Theta}$, the intended channel matrix is *proportional* to $\check{\boldsymbol{H}}_0$:

$$\check{\boldsymbol{H}} = \check{\boldsymbol{H}}_0 \boldsymbol{C}, \tag{4.44}$$

where $\boldsymbol{C}$ is a $2N_t \times 2N_t$ invertible matrix. We will next present a method to find the matrix $\boldsymbol{C}$.

4.3.2 Resolving the Ambiguity Matrix C

Let $\boldsymbol{H}_0$ denote the block Toeplitz matrix in the form of (4.32) obtained from the estimated matrix $\check{\boldsymbol{H}}_0$. Using (4.38), the received vector in (4.34) is reformulated as:

$$\boldsymbol{y}_c = \boldsymbol{H}_0 \begin{pmatrix} \boldsymbol{C} & & & \\ & \boldsymbol{C} & & \\ & & \ddots & \\ & & & \boldsymbol{C} \end{pmatrix} \boldsymbol{u} + \boldsymbol{w}. \tag{4.45}$$

By multiplying the received signal by the pseudo-inverse of $\boldsymbol{H}_0$, the modified $2N_tN \times 1$ received signal is given by:

$$\overline{\boldsymbol{y}}_c = \begin{pmatrix} \boldsymbol{C} & & & \\ & \boldsymbol{C} & & \\ & & \ddots & \\ & & & \boldsymbol{C} \end{pmatrix} \boldsymbol{u} + \overline{\boldsymbol{w}}, \tag{4.46}$$

where[6] $\overline{\boldsymbol{w}} = \boldsymbol{H}_0^{\#}\boldsymbol{w}$. By dividing the vector $\overline{\boldsymbol{y}}_c$ into N vectors of size $2N_t \times 1$:

$$\overline{\boldsymbol{y}}_c = [\overline{\boldsymbol{y}}_c^T(0), \overline{\boldsymbol{y}}_c^T(1), \ldots, \overline{\boldsymbol{y}}_c^T(N-1)]^T, \tag{4.47}$$

we have:

$$\overline{\boldsymbol{y}}_c(n) = \boldsymbol{C} \begin{pmatrix} \boldsymbol{x}(n) \\ \boldsymbol{s}(n) \end{pmatrix} + \overline{\boldsymbol{w}}(n),\ n = 0, \ldots, N-1. \tag{4.48}$$

From its definition, the matrix $\check{\boldsymbol{H}}$ is composed from the concatenation of two matrices, $\check{\boldsymbol{H}}^{rsi}$ and $\check{\boldsymbol{H}}^{s}$, representing the residual SI channel and the intended channel,

[6] $\boldsymbol{M}^{\#}$ denotes the pseudo-inverse of a given matrix $\boldsymbol{M}$.

respectively (i.e., $\check{\boldsymbol{H}} = \left[\check{\boldsymbol{H}}^{rsi}\check{\boldsymbol{H}}^{s}\right]$). In the same way, we divide $\boldsymbol{C}$ in two $2N_t \times N_t$ matrices $\boldsymbol{C}^i$ and $\boldsymbol{C}^s$ where the first one is associated with the SI and the second one is associated with the intended signal. Considering this division, we expand (4.48) as follows:

$$\overline{\boldsymbol{y}}_c(n) = \boldsymbol{C}^i\boldsymbol{x}(n) + \boldsymbol{C}^s s(n) + \overline{\boldsymbol{w}}(n),\ n = 0, \ldots, N-1, \tag{4.49}$$

or, by developing $\boldsymbol{x}(n) = \boldsymbol{x}_i(n) + \boldsymbol{G}\boldsymbol{x}_i^*(n) + \boldsymbol{A}_3\boldsymbol{x}_{ip3}(n)$ with $\boldsymbol{G} = \text{diag}\{g_{2,1}, \ldots, g_{2,N_t}\}$ and $\boldsymbol{A}_3 = \text{diag}\{\alpha_{3,1}, \ldots, \alpha_{3,N_t}\}$:

$$\overline{\boldsymbol{y}}_c(n) = \boldsymbol{C}^i\boldsymbol{x}_i(n) + \boldsymbol{C}^{iq}\boldsymbol{x}_i^*(n) + \boldsymbol{C}^{ip3}\boldsymbol{x}_{ip3}(n) + \boldsymbol{C}^s\boldsymbol{s}(n) + \overline{\boldsymbol{w}}(n), \tag{4.50}$$

where $\boldsymbol{x}_i(n) = [x_1(n), \ldots, x_{N_t}(n)]^T$, $\boldsymbol{x}_{ip3}(n) = \left[x_{1,ip3}(n), \ldots, x_{N_t,ip3}(n)\right]^T$, $\boldsymbol{C}^{iq} = \boldsymbol{C}^i\boldsymbol{G}$ and $\boldsymbol{C}^{ip3} = \boldsymbol{C}^i\boldsymbol{A}_3$. In (4.50), the vector $\boldsymbol{x}_i(n)$ is the undistorted SI while $\boldsymbol{C}^{iq}$ and $\boldsymbol{C}^{ip3}$ cover the effects of the IQ mixer and the PA, respectively. The vector $\overline{\boldsymbol{y}}_c(n)$ is the sum of a deterministic term representing the known transmitted self-signal, a stochastic term containing the intended signal received from the intended transmitter, and the additive noise. For a large number of subcarriers, the elements of the vector $\boldsymbol{s}(n)$ approach a Gaussian distribution [24]. Thus, we can reasonably assume that the unknown transmit symbols $\boldsymbol{s}(n)$ are Gaussian variables. Therefore, knowing the transmit vectors $\boldsymbol{x}_i(n)$ and $\boldsymbol{x}_{ip3}(n)$ and conditioned on the matrix $\boldsymbol{C}^s$, $\overline{\boldsymbol{y}}_c(N)$ is a Gaussian vector with mean $\boldsymbol{C}^i\boldsymbol{x}_i(n) + \boldsymbol{C}^{iq}\boldsymbol{x}_i^*(n) + \boldsymbol{C}^{ip3}\boldsymbol{x}_{ip3}(n)$ and covariance matrix $\boldsymbol{P} = \boldsymbol{C}^s\boldsymbol{R}_s\boldsymbol{C}^{sH} + \sigma^2\left(\check{\boldsymbol{H}}_0^H\check{\boldsymbol{H}}_0\right)^{-1}$. Adopting the Gaussian hypothesis, the log-likelihood function is given by:

$$\mathscr{L}\left(\boldsymbol{C}^i, \boldsymbol{C}^{iq}, \boldsymbol{C}^{ip3}, \boldsymbol{C}^s\right) = -N\log|\boldsymbol{P}| - \sum_{n=0}^{N-1}\left(\overline{\boldsymbol{y}}_c(n) - \boldsymbol{C}^i\boldsymbol{x}_i(n) - \boldsymbol{C}^{iq}\boldsymbol{x}_i^*(n) - \boldsymbol{C}^{ip3}\boldsymbol{x}_{ip3}(n)\right)^H\boldsymbol{P}^{-1}$$
$$\times\left(\overline{\boldsymbol{y}}_c(n) - \boldsymbol{C}^i\boldsymbol{x}_i(n) - \boldsymbol{C}^{iq}\boldsymbol{x}_i^*(n) - \boldsymbol{C}^{ip3}\boldsymbol{x}_{ip3}(n)\right), \tag{4.51}$$

where $|\cdot|$ returns the determinant of a matrix. The Maximum-Likelihood (ML) estimates of $\boldsymbol{C}^i$, $\boldsymbol{C}^{iq}$, $\boldsymbol{C}^{ip3}$ and $\boldsymbol{C}^s$ maximize the function $\mathscr{L}(\cdot)$ given in (4.51). The direct maximization of the cost function $\mathscr{L}(\cdot)$ requires a $6N_t^2$-dimensional grid search, which is intractable in practice. To overcome this complexity, we look to a closed-form expression of the solution.

Noting that $\mathscr{L}(\cdot)$ is a separable function of the matrices to estimate, we first minimize the cost function with respect to one matrix. The obtained minimum is a function of the other matrices. Then, we introduce this minimum back in the expression of the cost function to reduce the number of unknown. Minimizing this new function yields the global maximum of the original log-likelihood function [25].

We first maximize the log-likelihood function in (4.51) with respect to $\boldsymbol{P}$. The solution of this problem is [26]:

$$\boldsymbol{P}_{ML} = \frac{1}{N}\sum_{n=0}^{N-1}\left(\bar{\boldsymbol{y}}_c(n) - \boldsymbol{C}^i\boldsymbol{x}_i(n) - \boldsymbol{C}^{iq}\boldsymbol{x}_i^*(n) - \boldsymbol{C}^{ip3}\boldsymbol{x}_{ip3}(n)\right) \times$$
$$\left(\bar{\boldsymbol{y}}_c(n) - \boldsymbol{C}^i\boldsymbol{x}_i(n) - \boldsymbol{C}^{iq}\boldsymbol{x}_i^*(n) - \boldsymbol{C}^{ip3}\boldsymbol{x}_{ip3}(n)\right)^H. \quad (4.52)$$

Substituting $\boldsymbol{P}$ by $\boldsymbol{P}_{ML}$ into the log-likelihood function in (4.51), we obtain the so-called compressed likelihood function, that depends on the unknown matrices $\boldsymbol{C}^i$ $\boldsymbol{C}^{iq}$ and $\boldsymbol{C}^{ip3}$:

$$\mathcal{L}_c\left(\boldsymbol{C}^i, \boldsymbol{C}^{iq}, \boldsymbol{C}^{ip3}\right) = -\log\left|\sum_{n=0}^{N-1}\left(\bar{\boldsymbol{y}}_c(n) - \boldsymbol{C}^i\boldsymbol{x}_i(n) - \boldsymbol{C}^{iq}\boldsymbol{x}_i^*(n) - \boldsymbol{C}^{ip3}\boldsymbol{x}_{ip3}(n)\right) \times\right.$$
$$\left.\left(\bar{\boldsymbol{y}}_c(n) - \boldsymbol{C}^i\boldsymbol{x}_i(n) - \boldsymbol{C}^{iq}\boldsymbol{x}_i^*(n) - \boldsymbol{C}^{ip3}\boldsymbol{x}_{ip3}(n)\right)^H\right|, \quad (4.53)$$

where the terms irrelevant for the estimation have been discarded. The ML estimates of these matrices are given by:

$$\boldsymbol{C}^i_{ML}, \boldsymbol{C}^{iq}_{ML}, \boldsymbol{C}^{ip3}_{ML} = \arg\max_{\boldsymbol{C}^i, \boldsymbol{C}^{iq}, \boldsymbol{C}^{ip3}} \mathcal{L}_c\left(\boldsymbol{C}^i, \boldsymbol{C}^{iq}, \boldsymbol{C}^{ip3}\right). \quad (4.54)$$

At this point, we need to introduce some definitions. Let $\widetilde{\boldsymbol{C}}^i$ denotes the $2N_t^2 \times 1$ vector obtained by stacking all the columns of $\boldsymbol{C}^{iT}$ on top of each other (i.e., $\widetilde{\boldsymbol{C}}^i = \mathrm{vec}(\boldsymbol{C}^{iT})$) and $\widetilde{\boldsymbol{x}}_i(n)$ be the $2N_t \times 2N_t^2$ matrix given by:

$$\widetilde{\boldsymbol{x}}_i(n) = \boldsymbol{I}_{2N_t} \otimes \boldsymbol{x}_i^T(n). \quad (4.55)$$

$\widetilde{\boldsymbol{C}}^{iq}$, $\widetilde{\boldsymbol{C}}^{ip3}$ and $\widetilde{\boldsymbol{x}}_{ip3}(n)$ are also defined in the same way. Using these notations, the maximization problem in (4.54) is alternatively expressed as:

$$\widetilde{\boldsymbol{C}}^i_{ML}, \widetilde{\boldsymbol{C}}^{iq}_{ML}, \widetilde{\boldsymbol{C}}^{ip3}_{ML} = \arg\min_{\widetilde{\boldsymbol{C}}^i, \widetilde{\boldsymbol{C}}^{iq}, \widetilde{\boldsymbol{C}}^{ip3}} \left|\sum_{n=0}^{N-1}\left(\bar{\boldsymbol{y}}_c(n) - \widetilde{\boldsymbol{x}}_i(n)\widetilde{\boldsymbol{C}}^i - \widetilde{\boldsymbol{x}}_i^*(n)\widetilde{\boldsymbol{C}}^{iq} - \widetilde{\boldsymbol{x}}_{ip3}(n)\widetilde{\boldsymbol{C}}^{ip3}\right) \times\right.$$
$$\left.\left(\bar{\boldsymbol{y}}_c(n) - \widetilde{\boldsymbol{x}}_i(n)\widetilde{\boldsymbol{C}}^i - \widetilde{\boldsymbol{x}}_i^*(n)\widetilde{\boldsymbol{C}}^{iq} - \widetilde{\boldsymbol{x}}_{ip3}(n)\widetilde{\boldsymbol{C}}^{ip3}\right)^H\right|. \quad (4.56)$$

This modified problem allows us to obtain the following simple LS solution [27]:

$$\begin{pmatrix} \widetilde{\boldsymbol{C}}^i_{LS} \\ \widetilde{\boldsymbol{C}}^{iq}_{LS} \\ \widetilde{\boldsymbol{C}}^{ip3}_{LS} \end{pmatrix} = \left(\sum_{n=0}^{N-1}\widetilde{\boldsymbol{x}}^H(n)\widetilde{\boldsymbol{x}}(n)\right)^{-1}\sum_{n=0}^{N-1}\widetilde{\boldsymbol{x}}^H(n)\bar{\boldsymbol{y}}_c(n). \quad (4.57)$$

where $\widetilde{\boldsymbol{x}}(n) = \left[\widetilde{\boldsymbol{x}}_i^T(n)\, \widetilde{\boldsymbol{x}}_i^{*T}(n)\, \widetilde{\boldsymbol{x}}_{ip3}^T(n)\right]^T$. Note that the elements of $\widetilde{\boldsymbol{x}}_{ip3}(n)$ come from the cascade of the IQ mixer and the PA, and, hence, contain the signal image due to the IQ mixer unbalance. To simplify the estimation process, we approximate the elements of[7] $\widetilde{\boldsymbol{x}}_{ip3}(n)$ by $x_q(n)|x_q(n)|^2$. Since we are interested in the ML estimate, we define $\boldsymbol{\xi}_{ML}$ as the difference between the ML and LS estimates:

$$\boldsymbol{\xi}_{ML} = \begin{pmatrix} \widetilde{\boldsymbol{C}}^i_{ML} \\ \widetilde{\boldsymbol{C}}^{iq}_{ML} \\ \widetilde{\boldsymbol{C}}^{ip3}_{ML} \end{pmatrix} - \begin{pmatrix} \widetilde{\boldsymbol{C}}^i_{LS} \\ \widetilde{\boldsymbol{C}}^{iq}_{LS} \\ \widetilde{\boldsymbol{C}}^{ip3}_{LS} \end{pmatrix}, \tag{4.58}$$

and let $\boldsymbol{\xi} = \begin{pmatrix} \widetilde{\boldsymbol{C}}^i \\ \widetilde{\boldsymbol{C}}^{iq} \\ \widetilde{\boldsymbol{C}}^{ip3} \end{pmatrix} - \begin{pmatrix} \widetilde{\boldsymbol{C}}^i_{LS} \\ \widetilde{\boldsymbol{C}}^{iq}_{LS} \\ \widetilde{\boldsymbol{C}}^{ip3}_{LS} \end{pmatrix}$ denote the difference between the LS solution and a given value of $\widetilde{\boldsymbol{C}}^i$, $\widetilde{\boldsymbol{C}}^{iq}$ and $\widetilde{\boldsymbol{C}}^{ip3}$. We also consider the following two notations:

$$\begin{aligned} \boldsymbol{d}(n) &= \bar{\boldsymbol{y}}_c(n) - \widetilde{\boldsymbol{x}}_i(n)\widetilde{\boldsymbol{C}}^i_{LS} - \widetilde{\boldsymbol{x}}_i^*(n)\widetilde{\boldsymbol{C}}^{iq}_{LS} - \widetilde{\boldsymbol{x}}_{ip3}(n)\widetilde{\boldsymbol{C}}^{ip3}_{LS}, \\ \widehat{\boldsymbol{R}}_d &= \frac{1}{N}\sum_{n=0}^{N-1} \boldsymbol{d}(n)\boldsymbol{d}^H(n). \end{aligned} \tag{4.59}$$

As shown in Appendix, the optimization problem at hand is equivalent to:

$$\boldsymbol{\xi}_{ML} = \arg\min_{\boldsymbol{\xi}} \sum_{n=0}^{N-1} \boldsymbol{\xi}^H \widetilde{\boldsymbol{x}}^H(n)\widehat{\boldsymbol{R}}_d^{-1}\widetilde{\boldsymbol{x}}(n)\boldsymbol{\xi} - \boldsymbol{d}^H(n)\widehat{\boldsymbol{R}}_d^{-1}\widetilde{\boldsymbol{x}}(n)\boldsymbol{\xi} - \boldsymbol{\xi}\widetilde{\boldsymbol{x}}^H(n)\widehat{\boldsymbol{R}}_d^{-1}\boldsymbol{d}(n). \tag{4.60}$$

Its solution is obtained by nulling the derivative with respect to $\boldsymbol{\xi}$:

$$\boldsymbol{\xi}_{ML} = \left(\sum_{n=0}^{N-1} \widetilde{\boldsymbol{x}}^H(n)\widehat{\boldsymbol{R}}_d^{-1}\widetilde{\boldsymbol{x}}(n)\right)^{-1} \sum_{n=0}^{N-1} \widetilde{\boldsymbol{x}}^H(n)\widehat{\boldsymbol{R}}_d^{-1}\boldsymbol{d}(n). \tag{4.61}$$

Rearranging the expression in (4.61) using the notations given above, the ML estimate is given by:

$$\begin{pmatrix} \widetilde{\boldsymbol{C}}^i_{ML} \\ \widetilde{\boldsymbol{C}}^{iq}_{ML} \\ \widetilde{\boldsymbol{C}}^{ip3}_{ML} \end{pmatrix} = \left(\sum_{n=0}^{N-1} \widetilde{\boldsymbol{x}}^H(n)\widehat{\boldsymbol{R}}_d^{-1}\widetilde{\boldsymbol{x}}(n)\right)^{-1} \sum_{n=0}^{N-1} \widetilde{\boldsymbol{x}}^H(n)\widehat{\boldsymbol{R}}_d^{-1}\bar{\boldsymbol{y}}_c(n), \tag{4.62}$$

Note that the difference between the ML and LS estimates comes from the term $\widehat{\boldsymbol{R}}_d^{-1}$ in (4.62).

---

[7]By this approximation, we ignore the amplitude of $\alpha_{3,q}g_{2,q}$ compared to $\alpha_{3,q}$ and $g_{2,q}$.

For completeness, we present a method to find the ambiguity matrix of the intended channel $\boldsymbol{C}^s$. Using the estimate in (4.62), we obtain a cleaner version of $\bar{\boldsymbol{y}}_c(n)$ as $\bar{z}(n) = \bar{\boldsymbol{y}}_c(n) - \widetilde{\boldsymbol{x}}_i(n)\widetilde{\boldsymbol{C}}^i_{ML} - \widetilde{\boldsymbol{x}}^*_i(n)\widetilde{\boldsymbol{C}}^{iq}_{ML} - \widetilde{\boldsymbol{x}}_{ip3}(n)\widetilde{\boldsymbol{C}}^{ip3}_{ML}$. Assuming that pilot symbols are inserted in the subcarriers indexed by $\mathscr{P} = \{p_1, \ldots, p_{P_{\text{pilot}}}\}$, then the intended transmitted signal at antenna q is the sum of:

$$
\begin{aligned}
s^p_q(n) &= \frac{1}{\sqrt{N}} \sum_{i=1}^{P_{\text{pilot}}} S_q(p_i) e^{j2\pi p_i n/N}, \\
s^d_q(n) &= \frac{1}{\sqrt{N}} \sum_{k \notin \mathscr{P}} S_q(k) e^{j2\pi kn/N},
\end{aligned} \tag{4.63}
$$

where the first sequence $s^p_q(n)$ contains the pilot symbols and the second sequence $s^d_q(n)$ contains the unknown data symbols. By separating the pilot and data sequences in the expression of $\bar{z}(n)$, $\boldsymbol{C}^s$ can be obtained as:

$$
\widehat{\widetilde{\boldsymbol{C}}}^s = \left(\sum_{n=0}^{N-1} \widetilde{\boldsymbol{s}}^{pH}(n) \widetilde{\boldsymbol{s}}^p(n) \right)^{-1} \sum_{n=0}^{N-1} \widetilde{\boldsymbol{s}}^{pH}(n) \bar{z}(n), \tag{4.64}
$$

where $\widetilde{\boldsymbol{s}}^p(n)$ is defined in the same way as $\widetilde{\boldsymbol{x}}_i(n)$ using the pilot sequence $s^p_q(n)$ instead of the *self* signal $x_q(n)$.

4.4 Illustrative Results

In this section, we provide some simulation results on the performance of the proposed cancellation schemes applied to a MIMO full-duplex system using OFDM-4QAM with $N = 64$. A complete transmission chain is implemented to model the PA, the IQ mixer, the LNA and the ADC. The PA is modeled by a memory polynomial whose coefficients are derived based on practical values of the intercept points. The image rejection ratio of the IQ mixer is set to 28 dB. The ADC is realized by a 14-bit uniform quantizer to incorporate the quantization noise. Therefore, most of the nonlinearities of the transceiver chain are modeled. These parameters are also used in the following chapters, unless specified otherwise. The wireless channels are represented by multipath fading models with nine paths (i.e., $L = 8$). The SI channels are measured while the intended channel taps are generated as complex zero-mean i.i.d. Gaussian random variables. The amounts on antenna isolation is 40 dB, which is a realistic number reported in a many previous implementation of full-duplex systems [10, 28]. In the following, the SNR is the average intended-signal-to-thermal noise power ratio. Unless specified otherwise, the received intended-signal-to-SI power ratio ($\text{SIR}_{\text{input}}$) at the received input (before cancellation) is assumed to be −50 dB (i.e., the SI is 50 dB higher than the intended signal).

We compare the performance of the proposed subspace algorithm with the LS estimator with two different scenarios for the LS estimator. In the first scenario, the intended signal is considered as additional noise and only the residual SI channel is estimated. In this case, the estimate $\boldsymbol{h}_{LS}^{rsi}$ of $\boldsymbol{h}^{rsi}$ is equal to $\boldsymbol{X}^{\#}\boldsymbol{y}$. In the second scenario, we assume an ideal case that the intended signal is also known and both the residual SI and intended channels are estimated as

$$\begin{pmatrix} \widehat{\boldsymbol{h}}_{LS}^{rsi} \\ \widehat{\boldsymbol{h}}_{LS}^{s} \end{pmatrix} = [\boldsymbol{X}\ \boldsymbol{S}]^{\#}\boldsymbol{y}. \tag{4.65}$$

While the perfect knowledge of the intended signal is not a practical assumption, this procedure is taken as a reference to evaluate the performance of the proposed method. For both scenarios, the LS estimate uses 50 OFDM symbols. We also compare the proposed method with the widely-linear estimator proposed in [29] and the general LS formulation given in (2.13). Note that the algorithm in [29] ignores the effects of the PA nonlinearities and does not incorporate the intended signal in the estimation process. Let SINR_{rf} and SINR_{bb} denote the average intended-signal-to-residual-SI-and-noise power ratios after the RF cancellation stage and after the baseband cancellation stage, respectively. The sample covariance matrix is obtained with $T = 50$, 70 or 100 OFDM symbols, and the intended channel is assumed to be unchanged during T. Figures 4.1 and 4.2 represent the relation between the input SNR and output SINR after different cancellation stages for ($N_t = 1$, $N_r = 2$) and ($N_t = 2$, $N_r = 4$), respectively. It can be seen that the proposed algorithm greatly outperforms the LS-based when the intended signal is considered as noise (i.e., shown by the curve labelled "noisy LS" in the first scenario). Actually, in this scenario, the estimation error from the LS algorithm is very high, and thus, instead of reducing the SI, it introduces additional error, which makes SINR_{bb} after the baseband cancellation stage even lower than the SINR_{rf} after the RF cancellation stage. This result confirms the need to jointly estimate the SI and intended channels in order to obtain good cancellation performance. The cancellation performance of the LS-based using known intended symbols (i.e., shown by the curve labelled "joint LS" in the *ideal* second scenario) is greatly improved and comparable with that of the proposed algorithm at low input SNR. However, the joint LS-based algorithm needs the transmission of training symbols from the intended transmitter to obtain a good estimate of the SI channel while the proposed algorithm does not, and hence is more bandwidth-efficient. Moreover, the joint LS performance saturates as the SNR increases. This saturation is caused by the transmitter impairments, which are not modeled by the joint LS estimator. At high SNR, the proposed algorithm offers a superior performance approaching the perfect cancellation performance, especially for increased T. The good estimation performance of the proposed algorithm can be explained by the fact that the estimator exploits the information bearing in the unknown data to find the subspace

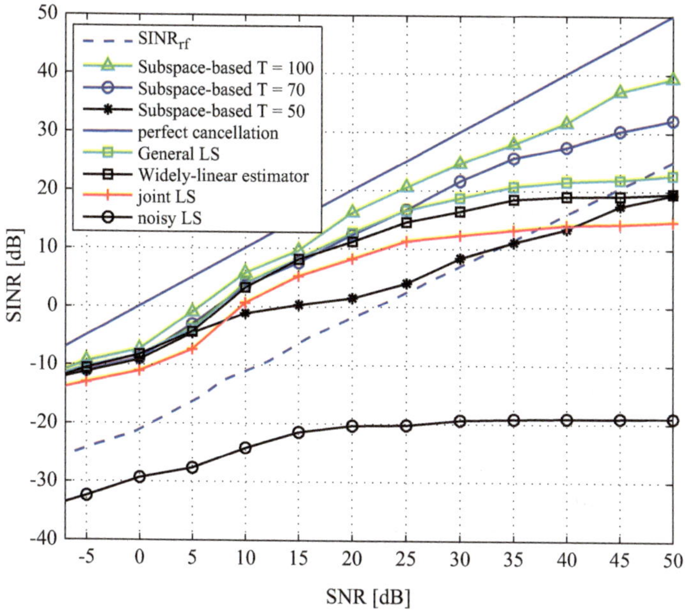

Fig. 4.1 Output SINR versus input SNR after different cancellation stages with $N_t = 1$ and $N_r = 2$

of the transmit signal and the remaining ambiguity factors are solved using the known SI data. On the other hand, the proposed method achieves performance close to that of the widely-linear estimator for SNR lower than 20 dB. At high SNR, the widely-linear estimator shows a noise floor because the PA nonlinearity is not considered during the estimation process while the general LS formulation includes the PA nonlinearity in the estimation process, which improves the SI cancellation at high SNR. Moreover, the proposed algorithm takes into account both the IQ imbalance and the PA nonlinearity. Besides, pilot frames incur an overhead and require synchronization between the two transceivers. This improvement comes at the cost of some computational complexity. Actually, the widely-linear algorithm involves the computation of the pseudo-inverse of a $TNN_r \times 2(L+1)N_r$ matrix. On the other hand, the proposed algorithm requires the eigen-decomposition of the $N_rM \times N_tM$ covariance matrix, the SVD of a $Np \times N_r(L+1)$ matrix (the matrix $\boldsymbol{\Theta}$ defined in (4.43) and the inverse of a $6N_t^2 \times 6N_t^2$ matrix to find the ambiguity terms in (4.62)). Table 4.1 summarizes the comparison of the widely-linear algorithm and the proposed subspace algorithm.

Figures 4.3 and 4.4 show the mean square error (MSE) of the obtained SI channel estimates versus the SNR for ($N_t = 1$, $N_r = 2$) and ($N_t = 2$, $N_r = 4$), respectively. As expected, the performance of the proposed algorithm is closely related to the accuracy of the sample covariance matrix, i.e., the MSE decreases with larger OFDM blocks. For 70 blocks, the corresponding MSE approaches that with

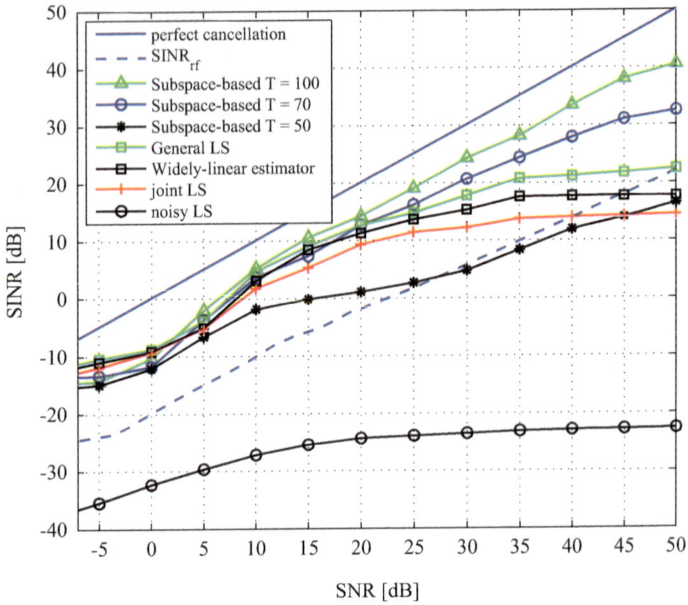

Fig. 4.2 Output SINR versus input SNR after different cancellation stages with $N_t = 2$ and $N_r = 4$

Table 4.1 Comparison of the proposed subspace algorithm and the widely-linear algorithm

	Subspace algorithm	Widely-linear algorithm in [29]
Advantages	– Joint estimation of the SI and intended channels – Reduces the PA nonlinearity – Does not require time-orthogonal periods	– Reduced complexity: requires the pseudo-inverse of a $TNN_r \times 2(L+1)N_r$ matrix
Disadvantages	Higher complexity: – Eigen-decomposition of a $N_rM \times N_rM$ matrix – SVD of a $Np \times N_r(L+1)p$ matrix – Inversion of a $6N_t^2 \times 6N_t^2$ matrix	– Requires time-orthogonal period – Does not reduce the PA nonlinearity – Does not estimate the intended channel

perfectly known covariance matrix. The widely-linear estimator still provides good estimation performance for low SNR but the presence of the PA nonlinearity acts as a noise floor which ultimately saturates its performance at high SNR. As it can be expected, the MSE performance of the LS-based estimation when considering the intended signal as noise is very poor. On the other hand, the LS-based joint estimation of the two channels presents relatively good performance at the expense of additional training sequence.

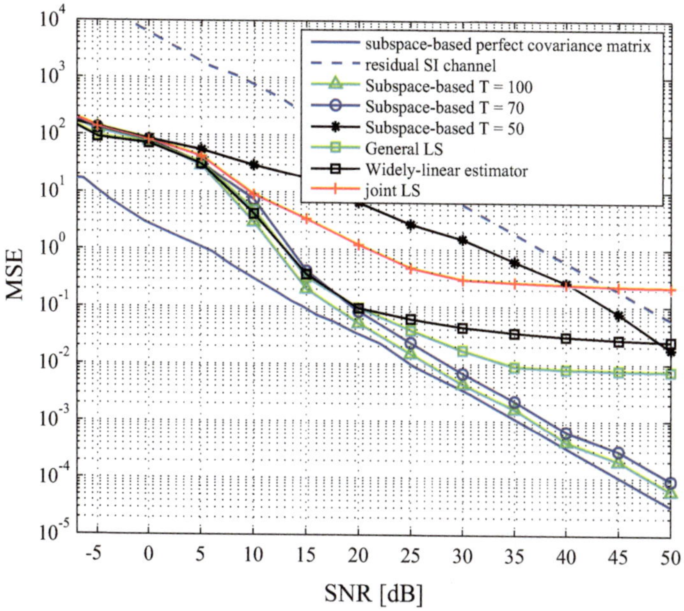

Fig. 4.3 Mean square error of the SI channel estimation versus input SNR for $N_t = 1$ and $N_r = 2$

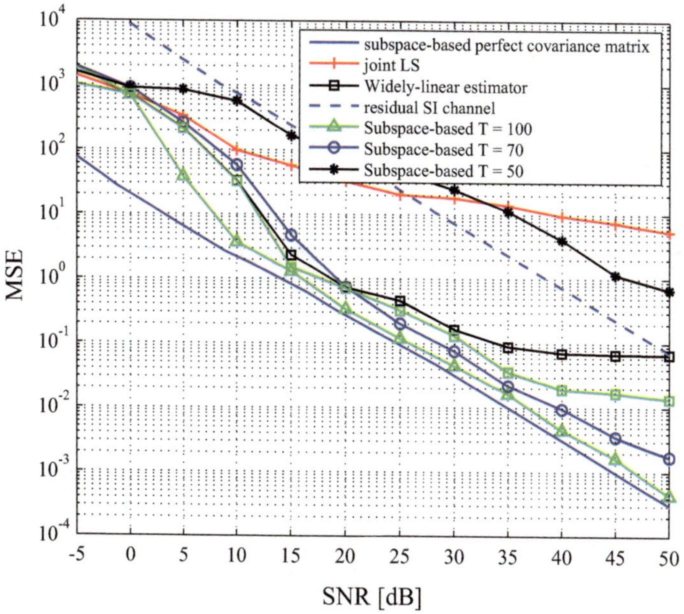

Fig. 4.4 Mean square error of the SI channel estimation versus input SNR for $N_t = 2$ and $N_r = 4$

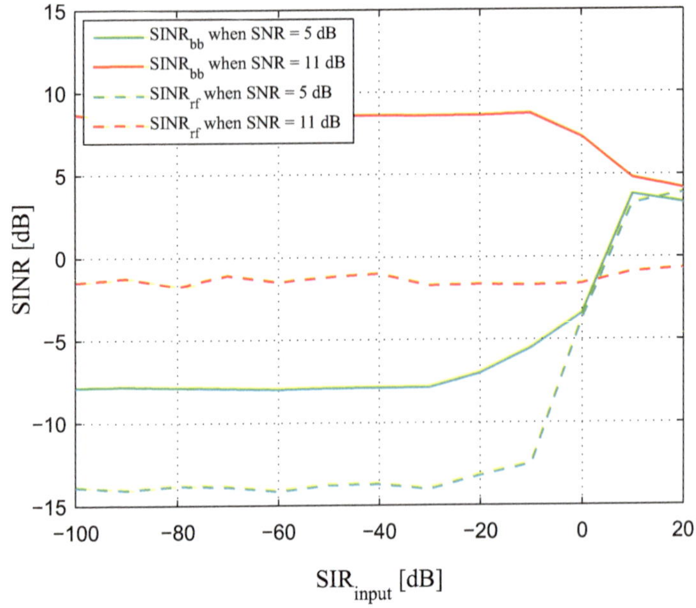

Fig. 4.5 Output SINR after different cancellation stages versus input SIR

In Figs. 4.1, 4.2, 4.3 and 4.4, as mentioned, the $\text{SIR}_{\text{input}}$ is fixed at -50 dB. Figure 4.5 shows the output SINR after different cancellation stages of the proposed algorithm for $\text{SIR}_{\text{input}}$ from -100 to 20 dB with SNR set to 11 and 5 dB. Clearly, SINR_{bb} remains constant over a wide range of $\text{SIR}_{\text{input}}$. For relatively high $\text{SIR}_{\text{input}}$ values, the SINR_{bb} after the baseband cancellation stage deteriorates as compared to SINR_{rf} after the RF cancellation stage. In this case, the SI is well reduced after the RF cancellation stage, which makes the channel estimation error in the baseband cancellation stage relatively high, and thus increases the residual SI.

To complete our study, performance curves are drawn for the MSE of the intended channel estimate in Fig. 4.6. As explained in Sect. 4.3.2, the subspace algorithm needs some known symbols to solve the signal channel ambiguity. To illustrate the impact of the training length, the LS-based estimator uses one OFDM symbol to estimate the channel and we vary the amount of known symbols in an OFDM block to be 50%, 25% and 16% of the OFDM block. Compared to the LS-based estimator, the proposed algorithm offers better performance using fewer known training symbols. Moreover, we obtain similar performance different pilot lengths. This shows that the subspace algorithm can robustly estimate the intended channel from few training symbols. Actually, using the subspace algorithm, the problem of estimating the $(L+1)N_rN_t$ channel coefficients is transformed to estimating the $2N_t^2 \times 1$ ambiguity vector $\widetilde{\boldsymbol{C}}^s$. Thus less parameters need to be estimated, which can be done from a reduced number of pilots.

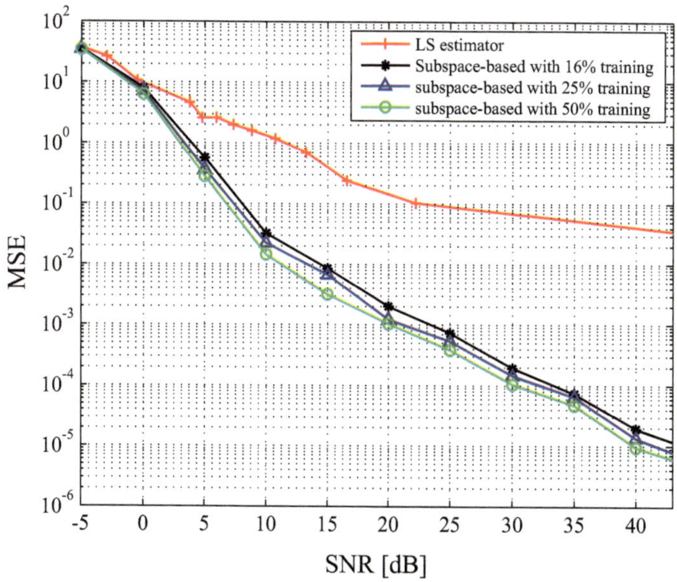

Fig. 4.6 Intended channel estimation MSE versus SNR

Figures 4.7 and 4.8 display, respectively, the bit error rate (BER) performance curves of the OFDM-4-QAM and OFDM-16-QAM systems using the proposed and LS estimators. In these figures, 25% and 16% of the transmit symbols are known to estimate the channel, represented by the dashed lines and the solid lines, respectively. These results show that the BER when using the LS estimators depends on the number of training symbols from the intended transceiver, while the proposed algorithm is not affected by the number of pilots and results in a significantly lower BER compared to the LS estimator. This is expected because the channel estimation error exhibits the similar tendency.

4.5 Chapter Summary

This chapter presented two estimation techniques for the RF and baseband SI-cancellation stages in full-duplex MIMO transceivers. The first algorithm for the RF SI-cancellation stage is based on the concept of CS to reduce the SI before the LNA. Then, in the baseband cancellation stage, a subspace-based estimator is applied to find the residual SI channel, the intended channel and to compensate the SI distortion caused by the transmitter impairments. The proposed algorithm performs a joint estimation of the different parameters by exploiting the available knowledge of the transmitted SI while the intended signal is unknown. Compared to the standard non-blind LS estimator, the proposed scheme does not require

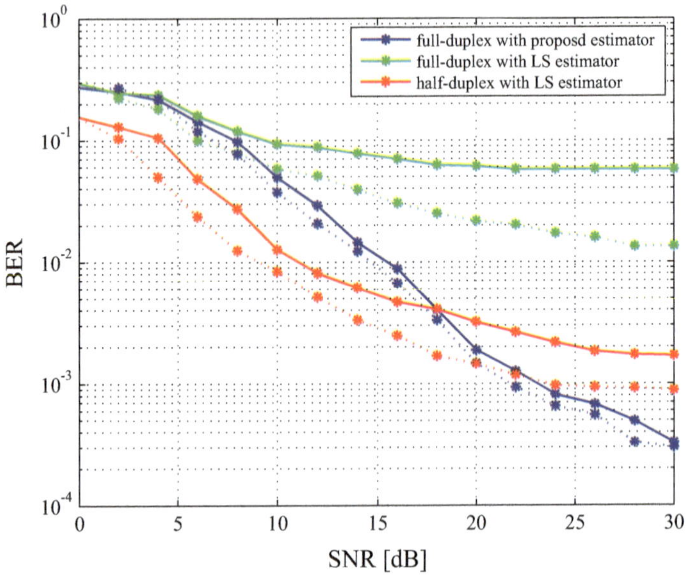

Fig. 4.7 Performance (BER) comparison of the proposed and LS estimators in full and half-duplex OFDM-4-QAM systems with 25% of pilots (*dashed lines*) and 16% of pilots (*solid lines*)

training blocks to find the residual SI channel and needs fewer training data to solve the intended channel ambiguity and, therefore, offers better bandwidth efficiency. Moreover, it is able to compensate the distorted SI. Simulation results have shown that the proposed algorithm improves the channel estimation accuracy and the cancellation performance.

Appendix: Proof of (4.60)

First, let $\widetilde{\boldsymbol{C}} = \begin{pmatrix} \widetilde{\boldsymbol{C}}^{i} \\ \widetilde{\boldsymbol{C}}^{iq} \\ \widetilde{\boldsymbol{C}}^{ip3} \end{pmatrix}$. Using the notations introduced in (4.58) and (4.59), we can write:

$$(\overline{\boldsymbol{y}}_c(n) - \widetilde{\boldsymbol{x}}(n)\widetilde{\boldsymbol{C}})(\overline{\boldsymbol{y}}_c(n) - \widetilde{\boldsymbol{x}}(n)\widetilde{\boldsymbol{C}})^H = (\overline{\boldsymbol{y}}_c(n) - \widetilde{\boldsymbol{x}}(n)\,(\widetilde{\boldsymbol{C}}_{LS} + \boldsymbol{\xi}))(\overline{\boldsymbol{y}}_c(n) - \widetilde{\boldsymbol{x}}(n)\,(\widetilde{\boldsymbol{C}} + \boldsymbol{\xi}))^H, \tag{4.66}$$

and further develop to obtain:

$$\boldsymbol{d}(n)\boldsymbol{d}^H(n) - \boldsymbol{d}(n)(\widetilde{\boldsymbol{x}}(n)\boldsymbol{\xi})^H - \widetilde{\boldsymbol{x}}(n)\boldsymbol{\xi}\boldsymbol{d}^H(n) + \widetilde{\boldsymbol{x}}(n)\boldsymbol{\xi}\boldsymbol{\xi}^H\widetilde{\boldsymbol{x}}^H(n). \tag{4.67}$$

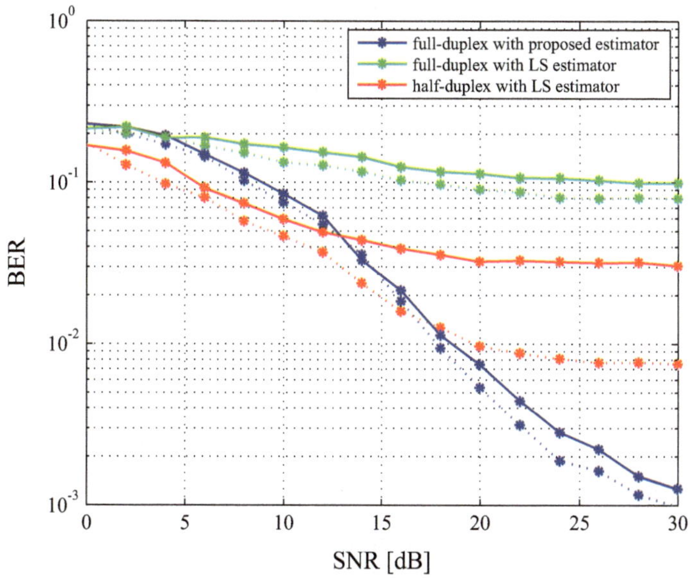

Fig. 4.8 Performance (BER) comparison of the proposed and LS estimators in full and half-duplex OFDM-16-QAM systems with 25% of pilots (*dashed lines*) and 16% of pilots (*solid lines*)

Injecting (4.67) into the cost function in (4.56), we obtain the following expression:

$$\left| \boldsymbol{R}_d + \frac{1}{N} \sum_{n=0}^{N-1} \boldsymbol{d}(n)(\widetilde{\boldsymbol{x}}(n)\boldsymbol{\xi})^H - \widetilde{\boldsymbol{x}}(n)\boldsymbol{\xi}\boldsymbol{d}^H(n) + \widetilde{\boldsymbol{x}}(n)\boldsymbol{\xi}\boldsymbol{\xi}^H\widetilde{\boldsymbol{x}}^H(n) \right|, \tag{4.68}$$

or the following equivalent cost function:

$$\left| \boldsymbol{I} + \frac{1}{N}\boldsymbol{R}_d^{-1} \sum_{n=0}^{N-1} \boldsymbol{d}(n)(\widetilde{\boldsymbol{x}}(n)\boldsymbol{\xi})^H - \widetilde{\boldsymbol{x}}(n)\boldsymbol{\xi}\boldsymbol{d}^H(n) + \widetilde{\boldsymbol{x}}(n)\boldsymbol{\xi}\boldsymbol{\xi}^H\widetilde{\boldsymbol{x}}^H(n) \right|. \tag{4.69}$$

Noting that, when N is large, the LS and ML estimates are close to the true value. Therefore, the vector $\boldsymbol{\xi}$ can be assumed small. Using the fact that, for[8] $||\boldsymbol{M}||_F << 1$, $|\boldsymbol{I} + \boldsymbol{M}| \approx 1 + \text{trace}(\boldsymbol{M})$ and the property that the trace is invariant under permutations, the minimization problem can be reduced to the one given in (4.60).

[8] $||\boldsymbol{M}||_F$ denotes the Frobenius norm of the matrix $\boldsymbol{M}$ [30].

References

1. M. Duarte, C. Dick, and A. Sabharwal, "Experiment-driven characterization of full-duplex wireless systems," *IEEE Trans. Wireless Comm.*, vol. 11, no. 12, pp. 4296–4307, 2012.
2. E. Candes and T. Tao, "Decoding by linear programming," *IEEE Trans. Inf. Theory*, vol. 51, no. 12, pp. 4203–4215, 2005.
3. E. Candes, J. Romberg, and T. Tao, "Stable signal recovery from incomplete and inaccurate measurements," *Comm. Pure appl. math.*, vol. 59, no. 8, pp. 1207–1223, 2006.
4. W. U. Bajwa, J. Haupt, A. M. Sayeed, and R. Nowak, "Compressed channel sensing: A new approach to estimating sparse multipath channels," *Proc. of the IEEE*, vol. 98, no. 6, pp. 1058–1076, 2010.
5. G. Tauböck and F. Hlawatsch, "Compressed sensing based estimation of doubly selective channels using a sparsity-optimized basis expansion," in *Proc. European Signal Processing Conf. (EUSIPCO'08)*, Lausanne, Switzerland, Aug. 2008.
6. A. Masmoudi and T. Le-Ngoc, "Subspace-based self-interference cancellation for full-duplex MIMO transceivers," *EURASIP Journal on Wireless Communications and Networking*, vol. 55, 2017. doi:10.1186/s13638-017-0839-x.
7. H. Krim and M. Viberg, "Two decades of array signal processing research: The parametric approach," *IEEE Signal Processing Magazine*, vol. 13, no. 4, pp. 67–94, 1996.
8. R. Schmidt, "Multiple emitter location and signal parameter estimation," *IEEE Trans. Antennas and Propagation*, vol. 34, no. 3, pp. 276–280, 1986.
9. A. Masmoudi and T. Le-Ngoc, "Self-interference cancellation for full-duplex MIMO transceivers," in *Proc. IEEE Wireless Commun. and Netw. Conf.*, New Orleans, Louisiana, USA, Mar. 2015.
10. J. I. Choi, M. Jain, K. Srinivasan, P. Levis, and S. Katti, "Achieving single channel, full duplex wireless communication," in *Proc. ACM MobiCom*, Chicago, Illinois, USA, Sept. 2010.
11. M. Duarte and A. Sabharwal, "Full-duplex wireless communications using off-the-shelf radios: Feasibility and first results," in *Proc. ASILOMAR Signals, Syst., Comput.*, Pacific Frove, California, USA, Nov. 2010.
12. M.-A. Baissas and A. M. Sayeed, "Pilot-based estimation of time-varying multipath channels for coherent CDMA receivers," *IEEE Trans. Signal Process.*, vol. 50, no. 8, pp. 2037–2049, 2002.
13. X. Ma, G. B. Giannakis, and S. Ohno, "Optimal training for block transmissions over doubly selective wireless fading channels," *IEEE Trans. Signal Process.*, vol. 51, no. 5, pp. 1351–1366, 2003.
14. H. Minn and N. Al-Dhahir, "Optimal training signals for MIMO OFDM channel estimation," *IEEE Trans. Wireless Comm.*, vol. 5, no. 5, pp. 1158–1168, 2006.
15. X. Wu, Y. Shen, and Y. Tang, "The power delay profile of the single-antenna full-duplex self-interference channel in indoor environments at 2.6 GHz," *IEEE Antennas and Wireless Propagation Letters*, vol. 13, pp. 1561–1564, 2014.
16. A. Sethi, V. Tapio, and M. Juntti, "Self-interference channel for full duplex transceivers," in *Proc. IEEE Wireless Commun. and Netw. Conf.*, Istanbul, Turkey, Apr. 2014.
17. C. R. Anderson, S. Krishnamoorthy, C. G. Ranson, T. J. Lemon, W. G. Newhall, T. Kummetz, and J. H. Reed, "Antenna isolation, wideband multipath propagation measurements, and interference mitigation for on-frequency repeaters," in *Proc. IEEE SoutheastCon*, Greensboro, North Carolina, USA, Mar. 2004.
18. S. S. Chen, D. L. Donoho, and M. A. Saunders, "Atomic decomposition by basis pursuit," *SIAM journal on scientific computing*, vol. 20, no. 1, pp. 33–61, 1998.
19. A. Masmoudi and T. Le-Ngoc, "Channel estimation and self-interference cancellation in full-duplex communication systems," *IEEE Trans. Vehicular Technology*, vol. 66, no. 1, pp. 321–334, Jan. 2017.
20. J. G. McMichael and K. E. Kolodziej, "Optimal tuning of analog self-interference cancellers for full-duplex wireless communication," in *50th Annual Allerton Conference on Communication, Control, and Computing*, Monticello, Illinois, USA, Oct. 2012.

21. J. Haupt, W. U. Bajwa, G. Raz, and R. Nowak, "Toeplitz compressed sensing matrices with applications to sparse channel estimation," *IEEE Trans. Inf. Theory*, vol. 56, no. 11, pp. 5862–5875, 2010.
22. A. Masmoudi and T. Le-Ngoc, "Self-interference cancellation limits in full-duplex communication systems," in *Proc. IEEE Global Telecommun. Conf.*, Washington DC, USA, Dec. 2016.
23. E. Moulines, P. Duhamel, J.-F. Cardoso, and S. Mayrargue, "Subspace methods for the blind identification of multichannel FIR filters," *IEEE Trans. Signal Process.*, vol. 43, no. 2, pp. 516–525, 1995.
24. H. Ochiai and H. Imai, "Performance analysis of deliberately clipped OFDM signals," *IEEE Trans. Comm.*, vol. 50, no. 1, pp. 89–101, 2002.
25. S. Talwar, M. Viberg, and A. Paulraj, "Blind separation of synchronous co-channel digital signals using an antenna array. Part I: Algorithms," *IEEE Trans. Signal Process.*, vol. 44, no. 5, pp. 1184–1197, 1996.
26. L. L. Scharf, *Statistical signal processing*. Addison-Wesley Reading, 1991, vol. 98.
27. S. M. Kay, *Fundamentals of statistical signal processing, Volume 1: Estimation theory*. Prentice Hall, 1993.
28. E. Everett, A. Sahai, and A. Sabharwal, "Passive self-interference suppression for full-duplex infrastructure nodes," *IEEE Trans. Wireless Comm.*, vol. 13, no. 2, pp. 680–694, 2014.
29. D. Korpi, L. Anttila, V. Syrjala, and M. Valkama, "Widely linear digital self-interference cancellation in direct-conversion full-duplex transceiver," *IEEE Journal on Selected Areas in Communications*, vol. 32, no. 9, pp. 1674–1687, Sept 2014.
30. G. H. Golub and C. F. Van Loan, *Matrix computations*. JHU Press, 2012, vol. 3.

Chapter 5
Widely-Linear Subspace-Based Self-Interference-Cancellation

Abstract This chapter addresses the self-interference (SI) cancellation at baseband for full-duplex MIMO communications systems in consideration of practical transmitter imperfections. In particular, we develop a subspace-based algorithm to jointly estimate the SI and intended channels and the nonlinear distortions. By exploiting the covariance and pseudo-covariance of the received signal, we can increase the dimension of the received signal subspace while keeping the dimension of the signal subspace constant, and hence, the proposed algorithm can be applied to most of full-duplex MIMO configurations with arbitrary numbers of transmit and receive antennas. The channel coefficients are estimated, up to an ambiguity term, without any knowledge of the intended signal. A joint detection and ambiguity identification scheme is proposed. Simulation results show that the proposed algorithm can properly estimate the channel with only one pilot symbol and offers superior SI cancellation performance.

Chapter 4 considers SI channel and parameter estimation for the RF and baseband cancellation stages. One notable aspect of the proposed subspace algorithm is the need of more receive antennas that transmit antennas to guarantee a nondegenerate noise subspace. This constraint may limit the application of the algorithm. In this chapter, we develop a subspace-based algorithm suitable for general MIMO full-duplex systems with arbitrary numbers of transmit and receive antennas. We exploit both the covariance and pseudo-covariance matrices of the received signal to effectively increase the dimension of the observation space while keeping the dimension of the signal subspace unchanged. The joint processing of the received signal and its complex conjugates, known as widely-linear processing, has been used in many works to improve the detection performance of various systems [1–3]. Also, in an entirely different context, the improper property of the received signal was first exploited for channel identification in [4] to obtain a virtual SIMO model from a SISO one. Other works follow in this direction for multiuser detection [5]. One recurrent assumption in these works is the use of real-valued symbols to obtain a non-zero pseudo-covariance matrix. We propose in this chapter a method to use the widely-linear processing to either real or complex symbols by forcing the transmit signal to be improper. We justify the advocated time domain approach and compare its performance to a frequency domain approach and we generalize the PA model to

T. Le-Ngoc, A. Masmoudi, *Full-Duplex Wireless Communications Systems*,
Wireless Networks, DOI 10.1007/978-3-319-57690-9_5

any nonlinearity order. As stated in Chap. 4, we cannot blindly recover the channel coefficients since an ambiguity term always appears in the final estimate [6–8]. This ambiguity is resolved using a sequence of pilot symbols, considerably shorter than needed in training-based techniques. In the following, we propose a joint data detection and estimation of the ambiguity term to considerably reduce the length of the pilot sequence. We show through simulation that just one pilot symbol is sufficient to perfectly estimate the channel.

This chapter is organized as follows. The subspace-based channel estimation is described in Sect. 5.1 with the widely-linear processing. In Sect. 5.2, we describe the joint decoding and ambiguity removal procedure. Illustrative simulation results are given in Sect. 5.3 and Sect. 5.4 presents the conclusion.

5.1 Widely-Linear Channel Estimator

We propose to apply a subspace-based algorithm to jointly estimate the SI and intended channel coefficients along with the nonlinear coefficients. The subspace method presented in Chap. 4 relies on the orthogonality property between the signal and noise subspaces. These two subspaces are obtained from eigendecomposition of the covariance matrix of the received signal[1] $\boldsymbol{y}$ after the RF cancellation stage. The covariance matrix $\boldsymbol{R}_y$ of the received vector $\boldsymbol{y}$ is given by:

$$\boldsymbol{R}_y = \boldsymbol{H}\boldsymbol{R}_u\boldsymbol{H}^H + \sigma^2\boldsymbol{I}_{MN_r}, \tag{5.1}$$

as long as the signal samples are uncorrelated from the noise samples.

The signal subspace is spanned by the columns of the matrix $\boldsymbol{H}$. Noting that the columns of $\boldsymbol{H}$ are, by construction, independent if there exists an $l \in [0,\ L]$ such that $\boldsymbol{H}(l)$ is full rank,[2] the matrix $\boldsymbol{H}$ is a full-rank matrix. Therefore, the dimension of the signal subspace is $2NN_t$. It follows that, to obtain a nondegenerate noise subspace, its dimension $N_rM - 2N_tN$ should be larger than zero, and thus, the number of receiving antennas should be larger than the number of transmitting antennas to make the subspace method work. In the particular case of $N_t = N_r$, the matrix $\boldsymbol{R}_y$ cannot be directly used to find the noise subspace. As an alternative, we consider the augmented received vector as:

$$\begin{aligned} \widetilde{\boldsymbol{y}} &= \begin{pmatrix} \boldsymbol{y} \\ \boldsymbol{y}^* \end{pmatrix} \\ &= \begin{pmatrix} \boldsymbol{H} & \boldsymbol{0} \\ \boldsymbol{0} & \boldsymbol{H}^* \end{pmatrix} \begin{pmatrix} \boldsymbol{u} \\ \boldsymbol{u}^* \end{pmatrix} + \begin{pmatrix} \boldsymbol{w} \\ \boldsymbol{w}^* \end{pmatrix}. \end{aligned} \tag{5.2}$$

[1] We mention that the received signal after the RF cancellation stage is simply referred by $\boldsymbol{y}_c$ in Chap. 4 and is referred by $\boldsymbol{y}$ in Chaps. 5 and 6.

[2] The previous condition is verified for independent channels between different antennas.

The use of the augmented received vector is usually referred as widely-linear processing. In this case, the augmented covariance matrix $\boldsymbol{R}_{\tilde{y}}$ of $\widetilde{\boldsymbol{y}}$ has the following structure:

$$\boldsymbol{R}_{\tilde{y}} = \widetilde{\boldsymbol{H}}\boldsymbol{R}_{\tilde{u}}\widetilde{\boldsymbol{H}}^H + \sigma^2\boldsymbol{I}_{2MN_r}, \tag{5.3}$$

where $\boldsymbol{R}_{\tilde{u}}$ denotes the covariance matrix of the augmented transmit signal $\widetilde{\boldsymbol{u}} = \begin{pmatrix} \boldsymbol{u} \\ \boldsymbol{u}^* \end{pmatrix}$ and:

$$\widetilde{\boldsymbol{H}} = \begin{pmatrix} \boldsymbol{H} & \boldsymbol{0} \\ \boldsymbol{0} & \boldsymbol{H}^* \end{pmatrix}. \tag{5.4}$$

It is worth mentioning that the proper noise has a vanishing pseudo-covariance [9–11]. The main purpose of using the extended received signal is to increase the dimension of the received signal and thus avoid the degenerate noise subspace. Hence, the subspace identification procedure can be derived only if the signal part covariance matrix, given by $\widetilde{\boldsymbol{H}}\boldsymbol{R}_{\tilde{u}}\widetilde{\boldsymbol{H}}^H$, of the covariance matrix $\boldsymbol{R}_{\tilde{y}}$ is singular. It results that $d_s = \mathrm{rank}(\widetilde{\boldsymbol{H}}\boldsymbol{R}_{\tilde{u}}\widetilde{\boldsymbol{H}}^H) < 2MN_r$. In this case, the signal is confined in a d_s-dimensional subspace and the remaining noise subspace is with dimension $2MN_r - d_s$. Singularity of $\boldsymbol{R}_{\tilde{u}}$ is a necessary condition to obtain a nondegenerate noise subspace. Actually, noting that $\widetilde{\boldsymbol{H}}$ is full-rank, non-singular $\boldsymbol{R}_{\tilde{u}}$ results in $\mathrm{rank}(\widetilde{\boldsymbol{H}}\boldsymbol{R}_{\tilde{u}}\widetilde{\boldsymbol{H}}^H) = 2MN_r$, and thus the matrix $\widetilde{\boldsymbol{H}}\boldsymbol{R}_{\tilde{u}}\widetilde{\boldsymbol{H}}^H$ spans all the observation space. On the other hand, since the matrix $\widetilde{\boldsymbol{H}}$ is a tall matrix, singularity of $\boldsymbol{R}_{\tilde{u}}$ is not a sufficient condition to guarantee the singularity of $\widetilde{\boldsymbol{H}}\boldsymbol{R}_{\tilde{u}}\widetilde{\boldsymbol{H}}^H$.

The matrix $\boldsymbol{R}_{\tilde{u}}$ can be expressed in a block form in terms of the covariance matrix of $\boldsymbol{u}$, $\boldsymbol{R}_u = \mathbb{E}\{\boldsymbol{u}\boldsymbol{u}^H\}$, the pseudo-covariance matrix $\boldsymbol{C}_u = \mathbb{E}\{\boldsymbol{u}\boldsymbol{u}^T\}$ and their complex conjugates as:

$$\boldsymbol{R}_{\tilde{u}} = \begin{pmatrix} \boldsymbol{R}_u & \boldsymbol{C}_u \\ \boldsymbol{C}_u^* & \boldsymbol{R}_u^* \end{pmatrix}. \tag{5.5}$$

In the following, we distinguish two cases of real and complex-modulated symbols.

5.1.1 Real-Modulated Symbols

For real-modulated symbols, it can be shown that $\boldsymbol{R}_{\tilde{u}} = \alpha^2\boldsymbol{M}\otimes\boldsymbol{I}_{2N_t}$ with the $2N\times 2N$ matrix $\boldsymbol{M}$ having the following form:

$$
\boldsymbol{M} = \left(\begin{array}{cccc|cccc}
1 & 0 & \ldots & 0 & 1 & 0 & \ldots & 0 \\
0 & \ddots & & & 0 & & & 1 \\
\vdots & & \ddots & & \vdots & & \iddots & \\
0 & \ldots & 0 & 1 & 0 & 1 & \ldots & 0 \\
\hline
1 & 0 & \ldots & 0 & 1 & 0 & \ldots & 0 \\
0 & & & 1 & 0 & \ddots & & \\
\vdots & & \iddots & & \vdots & & \ddots & \\
0 & 1 & \ldots & 0 & 0 & \ldots & 0 & 1
\end{array}\right), \tag{5.6}
$$

and α^2 is the variance of the transmitted signal. The block diagonal elements of $\boldsymbol{M}$ follow from the fact that $\mathbb{E}\{s_q(n)s_q^*(m)\} = \alpha^2$ if $n = m$ and 0 otherwise. Also, the non-diagonal blocks are obtained from:

$$
\begin{aligned}
\mathbb{E}\{s_q(n)s_q(m)\} &= \frac{1}{N}\sum_{p=0}^{N}\sum_{k=0}^{N}\mathbb{E}\{S_q(p)S_q(k)\}e^{\frac{-j2\pi}{N}(np+mk)} \\
&= \frac{1}{N}\sum_{p=0}^{N}\mathbb{E}\{S_q^2(p)\}e^{\frac{-j2\pi p}{N}(n+m)} \\
&= \begin{cases} \alpha^2, & \text{if } n+m=0 \text{ or } n+m=N, \\ 0, & \text{otherwise.} \end{cases}
\end{aligned} \tag{5.7}
$$

From (5.6), we note that each column of $\boldsymbol{M}$ appears exactly two times (the first column of $\boldsymbol{M}$ is the same as the $(N+1)$th column and the ith column of $\boldsymbol{M}$ is the same as the $(2N-i+2)$th column, for $i = 2, \ldots, N$). Therefore, the matrix $\boldsymbol{M}$ has exactly N independent columns and thus its rank is N. It follows that the rank of $\boldsymbol{R}_{\tilde{u}}$ is $2NN_t$.

In the following, we show that $\boldsymbol{R}_{\tilde{u}}$ has zero eigenvalue with multiplicity $2NN_t$ and $2\alpha^2$ also with multiplicity $2NN_t$. Since $\boldsymbol{M}$ is of rank N, then it has N strictly positive eigenvalues, $\tau_1, \tau_2, \ldots, \tau_N$, and eigenvalue 0 of multiplicity N. Since the covariance matrix $\boldsymbol{R}_{\tilde{u}}$ is given by $\alpha^2\boldsymbol{M} \otimes \boldsymbol{I}_{2N_t}$, it follows that $\boldsymbol{R}_{\tilde{u}}$ has also N eigenvalues $\tau_1, \tau_2, \ldots, \tau_N$ each of multiplicity $2N_t$ and eigenvalue 0 of multiplicity $2NN_t$. To find the non-zero eigenvalues, we solve the characteristic polynomial of $\boldsymbol{M}$ with respect to τ given by:

$$
\left|\boldsymbol{M} - \tau\boldsymbol{I}_{2N}\right| = 0, \tag{5.8}
$$

where $|.|$ returns the determinant of a matrix. First, if $\tau = 1$ is an eigenvalue of $\boldsymbol{M}$, then it exists a vector $\boldsymbol{a} \neq \boldsymbol{0}$ such that $\boldsymbol{Ma} - \boldsymbol{a} = \boldsymbol{0}$. It follows that $\boldsymbol{a}(1) = \boldsymbol{a}(2) =$

$\cdots = \boldsymbol{a}(2N) = 0$, which is in contradiction with $\boldsymbol{a} \neq \boldsymbol{0}$. Therefore, 1 is not an eigenvalue of $\boldsymbol{M}$.

By writing $\boldsymbol{M}$ as a block matrix:

$$\boldsymbol{M} = \begin{pmatrix} \boldsymbol{I}_N & \boldsymbol{M}_{1,2} \\ \boldsymbol{M}_{1,2} & \boldsymbol{I}_N \end{pmatrix}, \tag{5.9}$$

the characteristic polynomial of $\boldsymbol{M}$, for $\tau \neq 1$, is written as:

$$\begin{aligned} \left|\boldsymbol{M} - \tau \boldsymbol{I}_{2N}\right| &= \left|(1-\tau)\boldsymbol{I}_N\right|\left|(1-\tau)\boldsymbol{I}_N - \boldsymbol{M}_{1,2}(1-\tau)^{-1}\boldsymbol{I}_N\boldsymbol{M}_{1,2}\right| \\ &= (1-\tau)^N\left(1-\tau-(1-\tau)^{-1}\right)^N, \end{aligned} \tag{5.10}$$

where we used the fact that $\boldsymbol{M}_{1,2}\boldsymbol{M}_{1,2} = \boldsymbol{I}_N$. Then, the solutions to $\left|\boldsymbol{M} - \tau \boldsymbol{I}_{2N}\right| = 0$ are 0 and 2. Therefore, all non-zero eigenvalues of $\boldsymbol{M}$ are equal to 2 and thus all the non-zero eigenvalues of $\boldsymbol{R}_{\tilde{u}}$ are equal to $2\alpha^2$. Then, the matrix $\boldsymbol{R}_{\tilde{u}}$ is decomposed as $\boldsymbol{U}\boldsymbol{D}\boldsymbol{U}^H$ where $\boldsymbol{D}$ is the $4NN_t \times 4NN_t$ diagonal matrix with zeroes in the first $2NN_t$ diagonal elements and $2\alpha^2$ in the last $2NN_t$ diagonal elements and $\boldsymbol{U}$ is an orthogonal matrix whose columns are the corresponding eigenvectors of $\boldsymbol{R}_{\tilde{u}}$.

5.1.2 Complex-Modulated Symbols

For complex symbols, the pseudo-covariance matrix $\boldsymbol{C}_u$ is generally equal to the zero matrix, which makes the matrix $\boldsymbol{R}_{\tilde{u}}$ full rank. To avoid this problem, we apply a simple precoding at the input of the inverse fast Fourier transform (IFFT). It transforms the data symbol $\boldsymbol{X}_q$ to:

$$\widetilde{\boldsymbol{X}}_q = \boldsymbol{P}\boldsymbol{X}_q + \boldsymbol{Q}\boldsymbol{X}_q^*. \tag{5.11}$$

where $\boldsymbol{P}$ and $\boldsymbol{Q}$ are two matrices. By combining the data symbol $\boldsymbol{X}_q$ and its complex conjugate, we force the pseudo-covariance matrix to be different from zero.

The choice of $\boldsymbol{P}$ and $\boldsymbol{Q}$ can be done as follow. To make it simple, we consider the matrices $\boldsymbol{P}$ and $\boldsymbol{Q}$ having the following block structure:

$$\boldsymbol{P} = \begin{pmatrix} a\boldsymbol{I}_{N/2} & 0\boldsymbol{I}_{N/2} \\ 0\boldsymbol{I}_{N/2} & b\boldsymbol{I}_{N/2} \end{pmatrix},$$

$$\boldsymbol{Q} = \begin{pmatrix} 0\boldsymbol{I}_{N/2} & c\boldsymbol{I}_{N/2} \\ d\boldsymbol{I}_{N/2} & 0\boldsymbol{I}_{N/2} \end{pmatrix}, \tag{5.12}$$

for given real numbers a, b, c and d. Similarly to the real modulation, we have $\boldsymbol{R}_{\tilde{u}} = \alpha^2 \boldsymbol{M} \otimes \boldsymbol{I}_{2N_t}$ where $\boldsymbol{M}$ for complex modulation is given by:

$$\boldsymbol{M} = \begin{pmatrix} \boldsymbol{P}\boldsymbol{P}^T + \boldsymbol{Q}\boldsymbol{Q}^T & \boldsymbol{P}\boldsymbol{Q}^T + \boldsymbol{Q}\boldsymbol{P}^T \\ \boldsymbol{P}\boldsymbol{Q}^T + \boldsymbol{Q}\boldsymbol{P}^T & \boldsymbol{P}\boldsymbol{P}^T + \boldsymbol{Q}\boldsymbol{Q}^T \end{pmatrix}$$

$$= \begin{pmatrix} (a^2+c^2) & 0 & 0 & (ad+bc) \\ 0 & (b^2+d^2) & (ad+bc) & 0 \\ 0 & (ad+bc) & (a^2+c^2) & 0 \\ (ad+bc) & 0 & 0 & (b^2+d^2) \end{pmatrix} \otimes \boldsymbol{I}_{N/2},$$

for $a^2 + c^2 = b^2 + d^2$. Thus, for a, b, c and d satisfying $a^2 + c^2 = ad + bc$ and $b^2 + d^2 = ad + bc$, each line of $\boldsymbol{M}$ is repeated two times and $\boldsymbol{R}_{\tilde{u}}$ has rank $2NN_t$. As an example, we can take $a = 0.757$, $b = 0.5032$, $c = 0.4935$ and $d = 0.7506$. By doing so, the covariance matrix $\boldsymbol{R}_{\tilde{u}}$ has rank $2NN_t$ and can be decomposed as $\boldsymbol{U}\boldsymbol{D}\boldsymbol{U}^H$ with $\boldsymbol{D}$ the $4NN_t \times 4NN_t$ diagonal matrix with zeroes in the first $2NN_t$ diagonal elements.

5.1.3 Subspace-Based Algorithm

The noise subspace is the span of the $p = 2MN_r - 2NN_t$ eigenvectors of $\boldsymbol{R}_{\tilde{y}}$ corresponding to the smallest eigenvalue σ^2 and the columns of $\widetilde{\boldsymbol{H}}\boldsymbol{R}_{\tilde{u}}\widetilde{\boldsymbol{H}}^H$ belong to the signal subspace. Due to the orthogonality between the signal and the noise subspaces, each column of $\widetilde{\boldsymbol{H}}\boldsymbol{R}_{\tilde{u}}\widetilde{\boldsymbol{H}}^H$ is orthogonal to any vector in the noise subspace. Let $\{\boldsymbol{v}_i\}_{i=1}^p$ denote the p co-orthogonal eigenvectors corresponding to the smallest eigenvalue of $\boldsymbol{R}_{\tilde{y}}$. Then we have the following set of equations:

$$\boldsymbol{v}_i^H \widetilde{\boldsymbol{H}}\boldsymbol{R}_{\tilde{u}}\widetilde{\boldsymbol{H}}^H = \boldsymbol{0},\ i = 1, 2, \ldots, p. \tag{5.13}$$

From (5.13), we conclude that $\boldsymbol{v}_i$ spans the left null space of $\widetilde{\boldsymbol{H}}\boldsymbol{R}_{\tilde{u}}\widetilde{\boldsymbol{H}}^H$. For convenience, $\boldsymbol{U}$ is written as a block of four $2NN_t \times 2NN_t$ matrices:

$$\boldsymbol{U} = \begin{pmatrix} \boldsymbol{U}_1 & \boldsymbol{U}_2 \\ \boldsymbol{U}_3 & \boldsymbol{U}_4 \end{pmatrix}, \tag{5.14}$$

where the columns of $\left[\boldsymbol{U}_1^T, \boldsymbol{U}_3^T\right]^T$ are the eigenvectors of $\boldsymbol{R}_{\tilde{u}}$ corresponding to the eigenvalue zero and the columns of $\left[\boldsymbol{U}_2^T, \boldsymbol{U}_4^T\right]^T$ are the other eigenvectors. Then, taking into account the eigenvalue decomposition of $\boldsymbol{R}_{\tilde{u}}$, the set of equations in (5.13) are equivalent to:

$$\boldsymbol{v}_i^H \begin{pmatrix} \boldsymbol{H}\boldsymbol{U}_2 \\ \boldsymbol{H}^*\boldsymbol{U}_4 \end{pmatrix} = \boldsymbol{0},\ i = 1, 2, \ldots, p. \tag{5.15}$$

By dividing $\boldsymbol{v}_i$ into two $MN_r \times 1$ vectors, i.e., $\boldsymbol{v}_i = \left[\boldsymbol{v}_{i,1}^T,\ \boldsymbol{v}_{i,2}^T\right]^T$, (5.15) is rewritten as:

$$\boldsymbol{v}_{i,1}^H \boldsymbol{H} \boldsymbol{U}_2 + \boldsymbol{v}_{i,2}^H \boldsymbol{H}^* \boldsymbol{U}_4 = \boldsymbol{0}, \tag{5.16}$$

for $i = 1,\ 2, \ldots, p$. The matrix $\boldsymbol{H}$ is completely defined by the set of matrices $\boldsymbol{H}(l)$, for $l = 0,\ 1, \ldots,\ L$. Therefore, the specific structure of $\boldsymbol{H}$ should be taken into consideration when solving the equations in (5.16) to obtain a more accurate estimate of the channels. To that end, we divide the two vectors $\boldsymbol{v}_{i,1}$ and $\boldsymbol{v}_{i,2}$ as follows:

$$\boldsymbol{v}_{i,j} = \left[\boldsymbol{v}_{i,j}^T(1),\ \boldsymbol{v}_{i,j}^T(2), \ldots,\ \boldsymbol{v}_{i,j}^T(M)\right]^T,\ j = 1,\ 2,\ i = 1,\ 2, \ldots,\ p, \tag{5.17}$$

where each $\boldsymbol{v}_{i,j}(n)$, for $n = 1,\ 2, \ldots,\ M$, is a $N_r \times 1$ vector. From (4.32) and (5.17), each term $\boldsymbol{v}_{i,1}^H \boldsymbol{H}$ in (5.16) is rewritten as:

$$\begin{cases} \sum_{l=0}^{L} \boldsymbol{v}_{i,1}^H(n+l)\boldsymbol{H}(l), & n = N_{cp} - L + 1, \ldots, N - L \\ \sum_{l=0}^{\min(L, M-n)} \boldsymbol{v}_{i,1}^H(n+l)\boldsymbol{H}(l) + \sum_{l=\max(0, N-n+1)}^{L} \boldsymbol{v}_{i,1}^H(n-N+l)\boldsymbol{H}(l), & n = N+1-L, \ldots,\ M, \end{cases} \tag{5.18}$$

and $\boldsymbol{v}_{i,2}^H \boldsymbol{H}^*$ can also be partitioned in the same manner. By introducing $\check{\boldsymbol{h}}(l) = \text{vect}(\boldsymbol{H}(l))$ and $\boldsymbol{V}_{i,j}(n) = \boldsymbol{I}_{2N_t} \otimes \boldsymbol{v}_{i,j}^H(n)$, for $i = 1, \ldots,\ p$ and $j = 1,\ 2$, it is easy to verify that $\boldsymbol{v}_{i,j}^H(n)\boldsymbol{H}(l) = \check{\boldsymbol{h}}^T(l)\boldsymbol{V}_{i,j}^T(n)$. Let denote the $2NN_t \times 2N_tN_r(L+1)$ matrices $\boldsymbol{V}_{i,j}$, for $j = 1,\ 2$, as:

$$\boldsymbol{V}_{i,j} = \begin{pmatrix} \boldsymbol{V}_{i,j}(N_{cp} - L + 1) & \boldsymbol{V}_{i,j}(N_{cp} - L + 2) & \ldots & \boldsymbol{V}_{i,j}(N_{cp} + 1) \\ \boldsymbol{V}_{i,j}(N_{cp} - L + 2) & \boldsymbol{V}_{i,j}(N_{cp} - L + 3) & \ldots & \boldsymbol{V}_{i,j}(N_{cp} + 2) \\ \vdots & \vdots & \ddots & \vdots \\ \boldsymbol{V}_{i,j}(M - L) & \boldsymbol{V}_{i,j}(M - L + 1) & \ldots & \boldsymbol{V}_{i,j}(M) \\ \boldsymbol{V}_{i,j}(M - L + 1) & \ldots & \boldsymbol{V}_{i,j}(M) & \boldsymbol{0} \\ \vdots & \iddots & & \vdots \\ \boldsymbol{V}_{i,j}(M) & \boldsymbol{0} & & \boldsymbol{0} \end{pmatrix}$$

$$+\begin{pmatrix} \mathbf{0} & \cdots & \mathbf{0} \\ & \cdots & \vdots \\ \vdots & & \mathbf{0} \\ & & \boldsymbol{V}_{i,j}(1) \\ \mathbf{0} & \iddots & \vdots \\ \boldsymbol{V}_{i,j}(1) & \boldsymbol{V}_{i,j}(2) & \cdots & \boldsymbol{V}_{i,j}(L+1) \\ \boldsymbol{V}_{i,j}(2) & \boldsymbol{V}_{i,j}(3) & & \boldsymbol{V}_{i,j}(L+2) \\ \vdots & & & \vdots \\ \boldsymbol{V}_{i,j}(N_{cp}-L) & \boldsymbol{V}_{i,j}(N_{cp}-L+1) & \cdots & \boldsymbol{V}_{i,j}(N_{cp}) \end{pmatrix},$$

and $\check{\boldsymbol{h}} = [\check{\boldsymbol{h}}^T(0), \check{\boldsymbol{h}}^T(1), \ldots, \check{\boldsymbol{h}}^T(L)]^T$. Then, using the previous notations, (5.16) is rearranged to obtain:

$$\check{\boldsymbol{h}}^T \boldsymbol{V}_{i,1}^T \boldsymbol{U}_2 + \check{\boldsymbol{h}}^H \boldsymbol{V}_{i,2}^T \boldsymbol{U}_4 = \mathbf{0}, \tag{5.19}$$

or, by taking the transpose of the previous equation:

$$\boldsymbol{U}_2^T \boldsymbol{V}_{i,1} \check{\boldsymbol{h}} + \boldsymbol{U}_4^T \boldsymbol{V}_{i,2} \check{\boldsymbol{h}}^* = \mathbf{0}, \tag{5.20}$$

for $i = 1, 2, \ldots, p$. Note that the difference between (5.16) and (5.20) is that (5.20) takes into account the Toeplitz blocks structure of $\boldsymbol{H}$. Now, collecting all the previous equations, we obtain:

$$\widetilde{\boldsymbol{\Theta}}_1 \check{\boldsymbol{h}} + \widetilde{\boldsymbol{\Theta}}_2 \check{\boldsymbol{h}}^* = \mathbf{0}, \tag{5.21}$$

where:

$$\begin{aligned} \widetilde{\boldsymbol{\Theta}}_1 &= \left[(\boldsymbol{U}_2^T \boldsymbol{V}_{1,1})^T, (\boldsymbol{U}_2^T \boldsymbol{V}_{2,1})^T, \ldots, (\boldsymbol{U}_2^T \boldsymbol{V}_{p,1})^T\right]^T, \\ \widetilde{\boldsymbol{\Theta}}_2 &= \left[(\boldsymbol{U}_4^T \boldsymbol{V}_{1,2})^T, (\boldsymbol{U}_4^T \boldsymbol{V}_{2,2})^T, \ldots, (\boldsymbol{U}_4^T \boldsymbol{V}_{p,2})^T\right]^T. \end{aligned} \tag{5.22}$$

Separating the real and imaginary parts of (5.21), we have:

$$\underbrace{\begin{pmatrix} \Re(\widetilde{\boldsymbol{\Theta}}_1 + \widetilde{\boldsymbol{\Theta}}_2) & \Im(-\widetilde{\boldsymbol{\Theta}}_1 + \widetilde{\boldsymbol{\Theta}}_2) \\ \Im(\widetilde{\boldsymbol{\Theta}}_1 + \widetilde{\boldsymbol{\Theta}}_2) & \Re(\widetilde{\boldsymbol{\Theta}}_1 - \widetilde{\boldsymbol{\Theta}}_2) \end{pmatrix}}_{\overline{\boldsymbol{\Theta}}} \underbrace{\begin{pmatrix} \Re(\check{\boldsymbol{h}}) \\ \Im(\check{\boldsymbol{h}}) \end{pmatrix}}_{\overline{\boldsymbol{h}}} = \mathbf{0}. \tag{5.23}$$

From (5.23), the vector $\overline{\boldsymbol{h}}$ belongs to the right null space of $\overline{\boldsymbol{\Theta}}$. In practice, $\overline{\boldsymbol{h}}$ is a linear combination of the $4N_tN_r$ right singular vectors of the matrix $\overline{\boldsymbol{\Theta}}$, denoted by

$\boldsymbol{\beta}_i$, which are equal to the eigenvector of the Gramian $\overline{\boldsymbol{\Theta}\boldsymbol{\Theta}}^H$ corresponding to the zero eigenvalue. Therefore, an estimate of $\overline{\boldsymbol{h}}$ is given by:

$$\widehat{\overline{\boldsymbol{h}}} = \overline{\boldsymbol{\Phi}}\boldsymbol{c}, \tag{5.24}$$

where $\overline{\boldsymbol{\Phi}} = [\boldsymbol{\beta}_1, \boldsymbol{\beta}_2, \ldots, \boldsymbol{\beta}_{4N_tN_r}]$ and the $4N_tN_r \times 1$ vector $\boldsymbol{c}$ represents the ambiguity term to be estimated. The complex channel vector can also be obtained as:

$$\widehat{\check{\boldsymbol{h}}} = \boldsymbol{\Phi}\boldsymbol{c}, \tag{5.25}$$

where $\boldsymbol{\Phi}$ is obtained by combining the lines of $\overline{\boldsymbol{\Phi}}$ in the following way:

$$\overline{\boldsymbol{\Phi}} = \begin{pmatrix} \overline{\boldsymbol{\Phi}}_{real} \\ \overline{\boldsymbol{\Phi}}_{imag} \end{pmatrix} \rightarrow \boldsymbol{\Phi} = \overline{\boldsymbol{\Phi}}_{real} + j\overline{\boldsymbol{\Phi}}_{imag}, \tag{5.26}$$

and j is the complex number satisfying $j^2 = -1$.

We mention that the matrices $\boldsymbol{U}_2$ and $\boldsymbol{U}_4$ do not depend on the received signal and can be computed offline prior to the transmission. It is also seen that the over-estimated channel order L does not affect the estimation process. This is a common property with other subspace-based estimators [12].

5.2 Resolving the Ambiguity Term

As mentioned above, the subspace that contains the channels is obtained and the ambiguity term needs to be estimated to extract the exact coefficients. Different approaches can be applied to solve the ambiguity term $\boldsymbol{c}$. To do so, we highlight the contribution of $\boldsymbol{c}$ on the received vector $\boldsymbol{y}$. First, we separate the matrix $\boldsymbol{\Phi}$ in two $N_tN_r(L+1) \times 4N_tN_r$ matrices $\boldsymbol{\Phi}_i$ and $\boldsymbol{\Phi}_s$ which contribute in the SI and intended channels, respectively $\left(\text{i.e., } \check{\boldsymbol{h}}^{rsi} = \boldsymbol{\Phi}_i\boldsymbol{c} \text{ and } \check{\boldsymbol{h}}^{s} = \boldsymbol{\Phi}_s\boldsymbol{c}\right)$. By rearranging the elements of $\boldsymbol{\Phi}_i$ as:

$$\boldsymbol{\Phi}_i = \begin{pmatrix} \boldsymbol{\Phi}_{i,1}(0) \\ \boldsymbol{\Phi}_{i,2}(0) \\ \vdots \\ \boldsymbol{\Phi}_{i,N_t}(0) \\ \vdots \\ \boldsymbol{\Phi}_{i,1}(L) \\ \boldsymbol{\Phi}_{i,2}(L) \\ \vdots \\ \boldsymbol{\Phi}_{i,N_t}(L) \end{pmatrix} \rightarrow \check{\boldsymbol{\Phi}}_i = \begin{pmatrix} \boldsymbol{\Phi}_{i,1}(0) & \ldots & \boldsymbol{\Phi}_{i,N_t}(0) \\ \boldsymbol{\Phi}_{i,1}(1) & \ldots & \boldsymbol{\Phi}_{i,N_t}(1) \\ \vdots & & \vdots \\ \boldsymbol{\Phi}_{i,1}(L) & \ldots & \boldsymbol{\Phi}_{i,N_t}(L) \end{pmatrix}, \tag{5.27}$$

where each $\boldsymbol{\Phi}_{i,q}(l)$ is a $N_r \times 4N_tN_r$ matrix, $\check{\boldsymbol{H}}^{rsi} = \left[\boldsymbol{H}^{rsi^T}(0), \boldsymbol{H}^{rsi^T}(1), \ldots, \boldsymbol{H}^{rsi^T}(L)\right]^T$ can be written as:

$$\check{\boldsymbol{H}}^{rsi} = \check{\boldsymbol{\Phi}}_i(\boldsymbol{I}_{N_t} \otimes \boldsymbol{c}), \tag{5.28}$$

and $\check{\boldsymbol{H}}^s = \left[\boldsymbol{H}^{s^T}(0), \boldsymbol{H}^{s^T}(1), \ldots, \boldsymbol{H}^{s^T}(L)\right]^T$ can also be written as $\check{\boldsymbol{H}}^s = \check{\boldsymbol{\Phi}}_s(\boldsymbol{I}_{N_t} \otimes \boldsymbol{c})$, where $\check{\boldsymbol{\Phi}}_s$ is defined in the same way as $\check{\boldsymbol{\Phi}}_i$. $\check{\boldsymbol{H}}^{rsi}$ and $\check{\boldsymbol{\Phi}}_i$ are used to build the matrices $\boldsymbol{H}^{rsi}$ and $\boldsymbol{\Psi}_i$, respectively, having the same block structure as $\boldsymbol{H}$ in (4.32).

Next, we define the diagonal matrices $\boldsymbol{G}$ and $\boldsymbol{A}_p$ whose diagonal elements are, respectively, $\boldsymbol{g} = [g_{2,1}, \ldots, g_{2,N_t}]^T$ and $\boldsymbol{\alpha}_p = \left[\alpha_{2p+1,1}, \ldots, \alpha_{2p+1,N_t}\right]^T$, and we denote by $\boldsymbol{x}_{ip,p}(n) = \left[x_{1,ip,p}(n), \ldots, x_{N_t,ip,p}(n)\right]^T$, and $\boldsymbol{x}_{ip,p} = \left[\boldsymbol{x}_{ip,p}^T(0), \ldots, \boldsymbol{x}_{ip,p}^T(N-1)\right]^T$. Using the previous notations and by developing $\boldsymbol{x} = \boldsymbol{x}_i + (\boldsymbol{I}_N \otimes \boldsymbol{G})\boldsymbol{x}_i^* + \sum_{p=1}^{P} (\boldsymbol{I}_N \otimes \boldsymbol{A}_p)\boldsymbol{x}_{ip,p}$ in term of the transmitter impairments, one can express the received signal in (4.34) as:

$$\begin{aligned}
\boldsymbol{y} &= \underbrace{\boldsymbol{\Psi}_i(\boldsymbol{I}_{NN_t} \otimes \boldsymbol{c})}_{\boldsymbol{H}^{rsi}}\boldsymbol{x} + \underbrace{\boldsymbol{\Psi}_s(\boldsymbol{I}_{NN_t} \otimes \boldsymbol{c})}_{\boldsymbol{H}^{s}}\boldsymbol{s} + \boldsymbol{w}, \\
&= \boldsymbol{\Psi}_i(\boldsymbol{I}_{NN_t} \otimes \boldsymbol{c})\left(\boldsymbol{x}_i + (\boldsymbol{I}_N \otimes \boldsymbol{G})\boldsymbol{x}_i^* + \sum_{p=1}^{P}(\boldsymbol{I}_N \otimes \boldsymbol{A}_p)\boldsymbol{x}_{ip,p}\right) + \boldsymbol{\Psi}_s(\boldsymbol{I}_{NN_t} \otimes \boldsymbol{c})\boldsymbol{s} + \boldsymbol{w},
\end{aligned} \tag{5.29}$$

where $\boldsymbol{\Psi}_s$ and $\boldsymbol{H}^s$ are defined in the same way as $\boldsymbol{\Psi}_i$ and $\boldsymbol{H}^{rsi}$, respectively, and $\boldsymbol{s} = [\boldsymbol{s}^T(0), \ldots, \boldsymbol{s}^T(N-1)]^T$. After some manipulations, one can verify that $(\boldsymbol{I}_{NN_t} \otimes \boldsymbol{c})\boldsymbol{x}_i = (\boldsymbol{x}_i \otimes \boldsymbol{I}_{4N_tN_r})\boldsymbol{c}$ and $(\boldsymbol{I}_{NN_t} \otimes \boldsymbol{c})\boldsymbol{s} = (\boldsymbol{s} \otimes \boldsymbol{I}_{4N_tN_r})\boldsymbol{c}$. Then, the received vector in (5.29) is rewritten as:

$$\boldsymbol{y} = \boldsymbol{\Psi}_i\Big(\big(\boldsymbol{x}_i + (\boldsymbol{I}_N \otimes \boldsymbol{G})\boldsymbol{x}_i^* + \sum_{p=1}^{P}(\boldsymbol{I}_N \otimes \boldsymbol{A}_p)\boldsymbol{x}_{ip,p}\big) \otimes \boldsymbol{I}_{4N_tN_r}\Big)\boldsymbol{c} + \boldsymbol{\Psi}_s(\boldsymbol{s} \otimes \boldsymbol{I}_{4N_tN_r})\boldsymbol{c} + \boldsymbol{w}. \tag{5.30}$$

In (5.30), the received vector $\boldsymbol{y}$ is expressed as a linear function of the unknown vector $\boldsymbol{c}$. This formulation makes the estimation of $\boldsymbol{c}$ more tractable. While the transmitted SI is known, the distorted parts $(\boldsymbol{I}_N \otimes \boldsymbol{A}_p)\boldsymbol{x}_{ip,p}$ and $(\boldsymbol{I}_N \otimes \boldsymbol{G})\boldsymbol{x}_i^*$ of the SI from the cascade of the IQ mixer and PA need to be estimated. We begin by writing the following cost function $f(\boldsymbol{c}, \boldsymbol{s}, \boldsymbol{G}, \boldsymbol{A}_p) = ||\boldsymbol{y} - \boldsymbol{\Psi}_i((\boldsymbol{x}_i + (\boldsymbol{I}_N \otimes \boldsymbol{G})\boldsymbol{x}_i^* + \sum_{p=1}^{P}(\boldsymbol{I}_N \otimes \boldsymbol{A}_p)\boldsymbol{x}_{ip,p}) \otimes \boldsymbol{I}_{4N_tN_r})\boldsymbol{c} - \boldsymbol{\Psi}_s(\boldsymbol{s} \otimes \boldsymbol{I}_{4N_tN_r})\boldsymbol{c}||^2$ depending on $\boldsymbol{c}$, $\boldsymbol{G}$, $\boldsymbol{A}_p$ (for $p = 1, \ldots, P$) and $\boldsymbol{s}$. Given an initial estimate $\widehat{\boldsymbol{c}}$ of $\boldsymbol{c}$, the minimization of $f(\widehat{\boldsymbol{c}}, \boldsymbol{s}, \boldsymbol{G}, \boldsymbol{A}_p)$ with respect to $\boldsymbol{s}$, $\boldsymbol{G}$ and $\boldsymbol{A}_p$ can be recast as a LS problem. Then, using the solutions $\widehat{\boldsymbol{s}}$, $\widehat{\boldsymbol{G}}$ and $\widehat{\boldsymbol{A}}_p$, we minimize $f(\boldsymbol{c}, \widehat{\boldsymbol{s}}, \widehat{\boldsymbol{G}}, \widehat{\boldsymbol{A}}_p)$ with respect to $\boldsymbol{c}$. We iterate this procedure until the estimated parameters converge. An initial estimate of $\boldsymbol{c}$ is obtained using the LS criterion as:

$$\widehat{\boldsymbol{c}}_0 = \left(\boldsymbol{\Psi}_i\left(\boldsymbol{x}_i \otimes \boldsymbol{I}_{4N_tN_r}\right)\right)^{\#}\boldsymbol{y}, \tag{5.31}$$

where the operator $(\cdot)^{\#}$ returns the pseudo-inverse of a given matrix. At the kth iteration, the estimate $\widehat{\boldsymbol{c}}_{k-1}$ obtained at the previous iteration is used to find $\boldsymbol{s}$, $\boldsymbol{G}$ and $\boldsymbol{A}_p$ (or equivalently $\boldsymbol{g}$ and $\boldsymbol{\alpha}_p$) as follows:

$$\begin{pmatrix} \widehat{\boldsymbol{s}}_k \\ \widehat{\boldsymbol{g}}_k \\ \widehat{\boldsymbol{\alpha}}_{1,k} \\ \vdots \\ \widehat{\boldsymbol{\alpha}}_{P,k} \end{pmatrix} = \Big[\boldsymbol{\Psi}_s\widehat{\boldsymbol{C}}_{k-1},\ \boldsymbol{\Psi}_i\Big(\mathrm{diag}\{\boldsymbol{x}_i^*\}\boldsymbol{B}\Big) \otimes \widehat{\boldsymbol{c}}_{k-1},\ \boldsymbol{\Psi}_i\Big(\mathrm{diag}\{\boldsymbol{x}_{ip,1}\}\boldsymbol{B}\Big) \otimes \widehat{\boldsymbol{c}}_{k-1}, \ldots,$$

$$\boldsymbol{\Psi}_i\Big(\mathrm{diag}\{\boldsymbol{x}_{ip,P}\}\boldsymbol{B}\Big) \otimes \widehat{\boldsymbol{c}}_{k-1}\Big]^{\#}\Big(\boldsymbol{y} - \boldsymbol{\Psi}_i\widehat{\boldsymbol{C}}_{k-1}\boldsymbol{x}_i\Big), \tag{5.32}$$

where, for clarity, we introduce $\boldsymbol{B} = \boldsymbol{1}_N \otimes \boldsymbol{I}_{N_t}$ and $\widehat{\boldsymbol{C}}_{k-1} = \boldsymbol{I}_{NN_t} \otimes \widehat{\boldsymbol{c}}_{k-1}$ and we use the equality $\Big(\big((\boldsymbol{I}_N \otimes \boldsymbol{G})\boldsymbol{x}_i^*\big) \otimes \boldsymbol{I}_{4N_tN_r}\Big)\boldsymbol{c} = \Big(\big(\mathrm{diag}\{\boldsymbol{x}_i^*\}\boldsymbol{B}\big) \otimes \boldsymbol{c}\Big)\boldsymbol{g}$. Then, $\widehat{\boldsymbol{s}}_k$ is transformed in the frequency-domain and each element of the frequency-domain vector is projected to its closest discrete constellation point. The obtained vector is converted back to the time-domain to obtain a better estimate $\widetilde{\boldsymbol{s}}_k$ of $\boldsymbol{s}$. Then, an update of $\boldsymbol{c}$ at iteration k is obtained as:

$$\widehat{\boldsymbol{c}}_k = \left(\boldsymbol{\Psi}_i\left(\left(\boldsymbol{x}_i + \Big(\boldsymbol{I}_N \otimes \widehat{\boldsymbol{G}}_k\Big) + \sum_{p=1}^{P}\Big(\boldsymbol{I}_N \otimes \widehat{\boldsymbol{A}}_{p,k}\Big)\boldsymbol{x}_{ip,p}\right) \otimes \boldsymbol{I}_{4N_tNr}\right) + \boldsymbol{\Psi}_s\big(\widetilde{\boldsymbol{s}}_k \otimes \boldsymbol{I}_{4N_tN_r}\big)\right)^{\#}\boldsymbol{y}. \tag{5.33}$$

If a set of P_{pilot} pilot symbols are available at subcarriers indexed by $\mathscr{P} = \{p_1, \ldots, p_{P_{\text{pilot}}}\}$, the intended transmit signal at antenna q can be represented as the sum of two signals:

$$s_q^p(n) = \sum_{i=1}^{P_{\text{pilot}}} S_q(p_i)e^{j2\pi p_i n/N}, \quad s_q^d(n) = \sum_{k \notin \mathscr{P}} S_q(k)e^{j2\pi kn/N}, \tag{5.34}$$

where the first sequence $s_q^p(n)$ contains the pilot symbols and the second sequence $s_q^d(n)$ contains the unknown data symbols transmitted by other intended transmitter. Then, the received vector in (5.30) is rearranged as follows:

$$\boldsymbol{y} = \boldsymbol{\Psi}_i\left(\left(\boldsymbol{x}_i + (\boldsymbol{I}_N \otimes \boldsymbol{G})\boldsymbol{x}_i^* + \sum_{p=1}^{P}(\boldsymbol{I}_N \otimes \boldsymbol{A}_p)\boldsymbol{x}_{ip,p}\right) \otimes \boldsymbol{I}_{4N_tN_r}\right)\boldsymbol{c}$$
$$+ \boldsymbol{\Psi}_s\left((\boldsymbol{s}^p + \boldsymbol{s}^d) \otimes \boldsymbol{I}_{4N_tN_r}\right)\boldsymbol{c} + \boldsymbol{w}, \qquad (5.35)$$

where $\boldsymbol{s}^p$ and $\boldsymbol{s}^d$ are constructed in the same way as $\boldsymbol{s}$ and contain the pilot symbols and unknown symbols, respectively. The initial estimate of $\boldsymbol{c}$ is modified to incorporate the pilot symbols as:

$$\widehat{\boldsymbol{c}}_0 = \left(\boldsymbol{\Psi}_i(\boldsymbol{x}_i \otimes \boldsymbol{I}_{4N_tNr}) + \boldsymbol{\Psi}_s(\boldsymbol{s}^p \otimes \boldsymbol{I}_{4N_tN_r})\right)^{\#}\boldsymbol{y}, \qquad (5.36)$$

and the estimates of $\boldsymbol{s}^d$, $\boldsymbol{G}$ and $\boldsymbol{A}_p$ at iteration k are given by:

$$\begin{pmatrix} \widehat{\boldsymbol{s}}_k^d \\ \widehat{\boldsymbol{g}}_k \\ \widehat{\boldsymbol{\alpha}}_{1,k} \\ \vdots \\ \widehat{\boldsymbol{\alpha}}_{P,k} \end{pmatrix} = \Big[\boldsymbol{\Psi}_s\widehat{\boldsymbol{C}}_{k-1},\ \boldsymbol{\Psi}_i\left(\mathrm{diag}\{\boldsymbol{x}_i^*\}\boldsymbol{B}\right) \otimes \widehat{\boldsymbol{c}}_{k-1},\ \boldsymbol{\Psi}_i\left(\mathrm{diag}\{\boldsymbol{x}_{ip,1}\}\boldsymbol{B}\right) \otimes \widehat{\boldsymbol{c}}_{k-1}, \ldots,$$
$$\boldsymbol{\Psi}_i\left(\mathrm{diag}\{\boldsymbol{x}_{ip,P}\}\boldsymbol{B}\right) \otimes \widehat{\boldsymbol{c}}_{k-1}\Big]^{\#}\left(\boldsymbol{y} - \boldsymbol{\Psi}_i\widehat{\boldsymbol{C}}_{k-1}\boldsymbol{x}_i - \boldsymbol{\Psi}_s\widehat{\boldsymbol{C}}_{k-1}\boldsymbol{s}^p\right). \qquad (5.37)$$

As before, $\widehat{\boldsymbol{s}}_k^d$ is converted to the frequency-domain, demodulated then transformed to the time-domain to obtain $\widetilde{\boldsymbol{s}}_k^d$. The updated estimate of $\boldsymbol{c}$ at iteration k is obtained as:

$$\widehat{\boldsymbol{c}}_k = \left(\boldsymbol{\Psi}_i((\boldsymbol{x}_i + (\boldsymbol{I}_N \otimes \widehat{\boldsymbol{G}}_k)\boldsymbol{x}_i^* + \sum_{p=1}^{P}(\boldsymbol{I}_N \otimes \widehat{\boldsymbol{A}}_{3,p})\boldsymbol{x}_{ip,p}) \otimes \boldsymbol{I}_{4N_tN_r}) + \boldsymbol{\Psi}_s((\boldsymbol{s}^p + \widetilde{\boldsymbol{s}}_k^d) \otimes \boldsymbol{I}_{4N_tN_r})\right)^{\#}\boldsymbol{y}. \qquad (5.38)$$

In the following, we summarize the different steps of the proposed algorithm:

1. Compute the augmented covariance matrix $\boldsymbol{R}_{\tilde{y}}$ by time averaging of T received samples as:

$$\widehat{\boldsymbol{R}}_{\tilde{y}} = \frac{1}{T}\sum_{t=1}^{T}\begin{pmatrix} \boldsymbol{y}_t \\ \boldsymbol{y}_t^* \end{pmatrix}\begin{pmatrix} \boldsymbol{y}_t \\ \boldsymbol{y}_t^* \end{pmatrix}^H$$

2. Perform eigendecomposition of $\boldsymbol{R}_{\tilde{y}}$ and take the p eigenvectors $\boldsymbol{v}_i$ corresponding to the smallest eigenvalue of $\boldsymbol{R}_{\tilde{y}}$.
3. Construct the matrix $\overline{\boldsymbol{\Theta}}$ from $\boldsymbol{v}_i$ and compute the $4N_tN_r$ singular vectors of $\overline{\boldsymbol{\Theta}}$ corresponding to the zero singular value to form $\overline{\boldsymbol{\Phi}}$.

4. Build the matrices $\check{\boldsymbol{\Phi}}_i$ and $\check{\boldsymbol{\Phi}}_s$ as given in (5.27).
5. Estimate the ambiguity vector $\boldsymbol{c}$ by iterating between (5.32) and (5.33) if no pilot symbols are available or between (5.37) and (5.38) if a set of pilot symbols are available from the intended transceiver.

Both algorithms in this chapter and Chap. 4 use the subspace concept to estimate the unknown parameters. Yet, both of them are applied to two different situations and the derivations and details included in this chapter are different from those disclosed in Chap. 4. In fact, although the objective is the same that the estimation problem is tackled via the subspace concept, the formulation and the application of the subspace technique are different from those disclosed in Chap. 4. In the following, we summarise the main different points in the two chapters. We refer to the algorithm presented in Chap. 4 by linear subspace.

- The linear subspace algorithm is developed under the assumption that the number of receive antennas (N_r) is double than the number of transmit antennas (N_t) and cannot be applied when $N_r = N_t$. As detailed in the beginning of Sect. 5.1, the use of the augmented received vector can solve this problem.
- In this chapter, we have to make the $4NN_t \times 4NN_t$ covariance matrix $\boldsymbol{R}_{\tilde{u}}$ in (5.5) with rank $2NN_t$. To that end, we distinguish two cases:

  1. For real-modulated symbols, a complete arguments proving that the $4NN_t \times 4NN_t$ matrix $\boldsymbol{R}_{\tilde{u}}$ is of rank $2NN_t$ is given in Sect. 5.1.1.
  2. For complex-modulated symbols, direct application of the algorithm is not possible as the pseudo-covariance matrix $\boldsymbol{C}_u$ is equal to zero making $\boldsymbol{R}_{\tilde{u}}$ full rank. To solve this problem, we apply a precoding technique as detailed in Sect. 5.1.2. This precoding technique can open other field to exploit the subspace technique in communication systems. Actually, most of the applications related to the subspace concept are limited to larger receive antennas than transmit antennas and only real-modulated symbol are used when combined with the widely-linear formulation.

- The manner we manipulate the covariance matrix is different. In the linear subspace algorithm, $\boldsymbol{R}_u$ is full rank while in this chapter $\boldsymbol{R}_{\tilde{u}}$ is not full rank. Also, the use of the received signal and its complex conjugate creates redundancy on the estimated parameters as they appear with their complex conjugates in the orthogonal equations. These two facts lead to additional manipulations from (5.13) to (5.26).
- The way we solve the ambiguity term in the two chapters is different. In the linear subspace algorithm, we were able to separate the received OFDM signal over the N_r antennas into N vectors of size $2N_t \times 1$, each one depends on the ambiguity term in a linear manner. Then the ambiguity term is obtained by average over the N vectors as given in (4.62) and (4.64). However, in this chapter, the received signal is not divided but we are able to separate the contribution of the ambiguity term and the different nonlinearity coefficients (namely $\boldsymbol{g}$ and $\boldsymbol{\alpha}_p$ for $p = 1, \ldots, P$). The developments that lead to this separability are presented

in from (5.27) to (5.30). Finally, the obtained estimates of the ambiguity term [in (5.33) and (5.38)] and the nonlinearity coefficients [in (5.32) and (5.37)] are different from the estimates obtained in the linear subspace algorithm where we combined the ambiguity term and the nonlinearity coefficients in one term.

5.3 Simulation Results

In this section, we provide some simulation results on the performance of the proposed estimation algorithm for a 2×2 MIMO full-duplex system. The transmitted bits are mapped to 4-QAM symbols, then passed through an OFDM modulator of length $N = 64$. The wireless channels are represented by multipath Rayleigh fading with five paths. Since the exact number of paths is supposed to be unknown, the algorithm is parameterized as if there are eight paths. In the following, the SNR is defined as the average intended-signal-to-thermal noise power ratio and the estimation MSE of $\boldsymbol{H}$ is MSE $= \mathbb{E}\left\{||\boldsymbol{H} - \widehat{\boldsymbol{H}}||^2\right\}$. To model the RF impairments, a complete transmission chain is simulated. The PA coefficients are derived from the intercept points by taking the IIP3 $=$ 20 dBm [13]. For the IQ mixer, the ratio between the direct signal and the image is set to 28 dB which is specified in 3GPP LTE specifications [14]. The ADC is modeled as a 14-bit quantizer to incorporate the quantization noise. Therefore, no simplifications are made regarding the different impairments. Antenna separation can attenuate the SI by 40 dB while the RF cancellation stage reduces the SI by 30 dB [15].

The proposed algorithm is compared to different channel estimators: the least square (LS) and the maximum likelihood (ML) algorithms. For the LS estimator, the channel coefficients are obtained using the known *self* signal and the pilot symbols in the intended signal. It simply considers the unknown symbols as additive noise. The ML estimate is obtained by maximizing the following cost function [16]:

$$\mathscr{L}(\boldsymbol{H}^{rsi}, \boldsymbol{H}^{s}) = \log|\boldsymbol{R}| - \left(\boldsymbol{y} - \boldsymbol{H}^{rsi}\boldsymbol{x} - \boldsymbol{H}^{s}\boldsymbol{s}^{p}\right)^{H} \boldsymbol{R}^{-1} \left(\boldsymbol{y} - \boldsymbol{H}^{rsi}\boldsymbol{x} - \boldsymbol{H}^{s}\boldsymbol{s}^{p}\right),$$

where $\boldsymbol{R} = \alpha^2 \boldsymbol{H}^{sH}\boldsymbol{H}^{s} + \sigma^2 \boldsymbol{I}_{N_rM}$. Here we anticipate on the ML estimator detailed in Chap. 6. The covariance matrix is obtained by averaging 60 OFDM blocks. Figures 5.1 and 5.2 plot the MSE versus SNR curves for the SI and intended channel estimations, respectively. In both figures, one pilot symbol, from the intended transceiver, is used to solve the ambiguity matrix. For comparison purpose, a perfect estimate of the ambiguity term $\boldsymbol{c}$ is obtained as $\boldsymbol{c}_{perfect} = \arg\min_{\boldsymbol{c}} ||\check{\boldsymbol{h}} - \boldsymbol{\Phi}\boldsymbol{c}||_2^2$ and the corresponding curves are labelled by clairvoyant subspace. It is seen that, when one pilot symbol is used in the ML and LS estimators, the proposed subspace algorithm offers notably lower MSE over a large SNR range. We also represent the performance of the ML and LS estimators when 20% of the transmit symbols are known (the pilot symbols are equally spaced within one OFDM symbol) while keeping one pilot symbol for the subspace method. In this case, the three algorithms give comparable performance at low SNR region with the

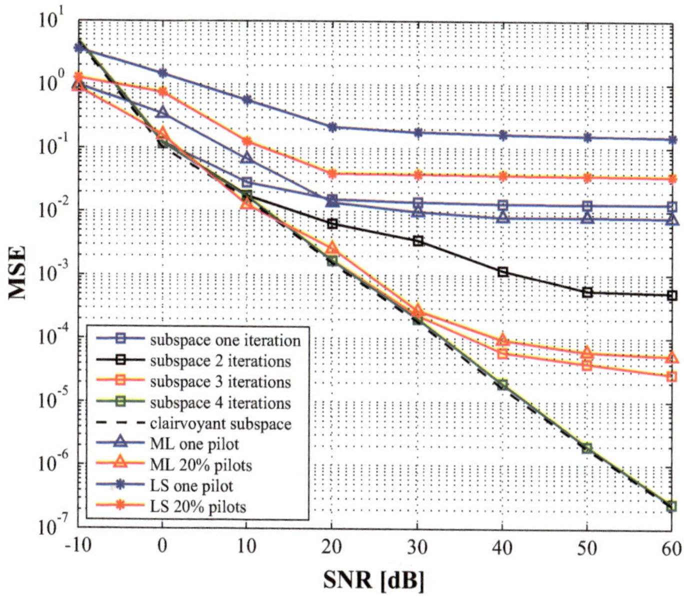

Fig. 5.1 SI channel estimation MSE versus SNR with 60 received OFDM symbols

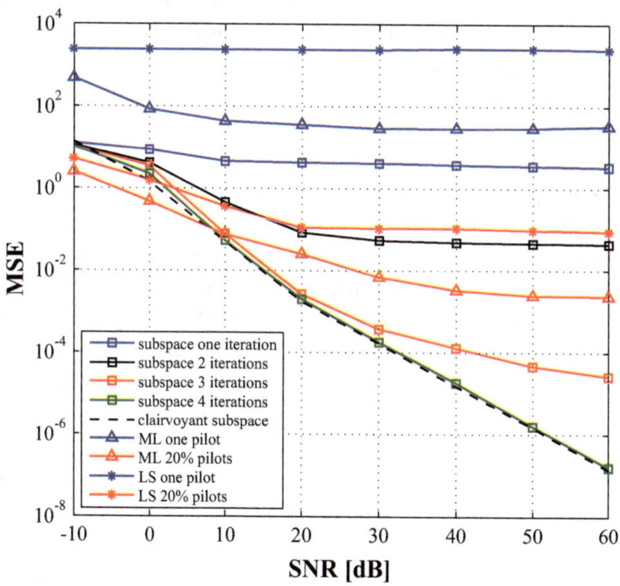

Fig. 5.2 Intended channel estimation MSE versus SNR with 60 received OFDM symbols

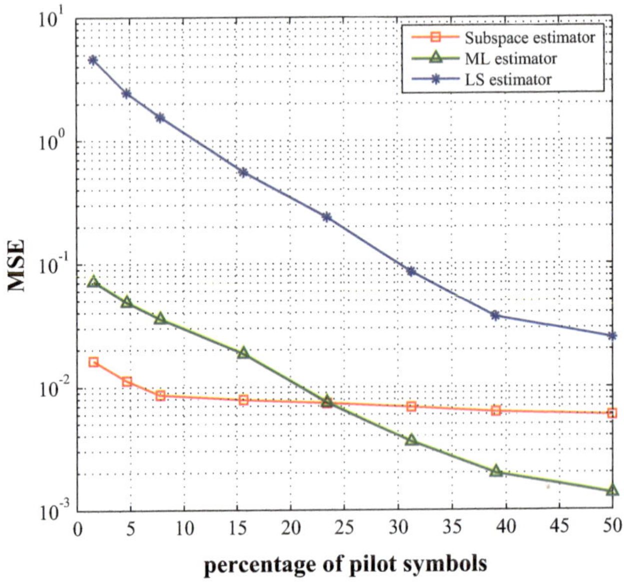

Fig. 5.3 SI channel estimation MSE versus percentage of pilot symbols for SNR = 10 dB

expense of lower bandwidth efficiency for the ML and LS algorithms. As the SNR increases, the performance of the LS estimator saturates due to the reduced number of pilot symbols and the presence of the unknown transmit signal from the intended transceiver which acts as an additive noise. While the subspace algorithm exploits the information bearing in the unknown data to find the signal subspace. The ambiguity term is first solved using the known transmit symbols, then the iterative decoding ambiguity estimation is applied to improve the estimation performance. From Figs. 5.1 and 5.2, three to four iterations are sufficient to converge and the obtained performance is close to the performance when the ambiguity term $\boldsymbol{c}$ is perfectly obtained. As it can be expected, the estimate of the SI channel is more accurate than the estimate of the intended channel. This can be explained by the fact that the self-signal is known while one pilot symbol is known in the intended signal.

The number of pilot symbols is a critical issue in channel estimation since a large pilot sequence provides better estimation performance but reduces the bandwidth efficiency of the system. In Figs. 5.3 and 5.4, we compare the impact of the number of pilot symbols on the performance of the three estimators. It can be seen from these figures that the subspace method is not greatly affected by the number of pilot symbols since the subspaces are obtained using the second-order statistics of the received signal and not the transmit signal itself. Clearly, the proposed algorithm outperforms the ML and LS estimators at a reduced number of pilots while this tendency is inverted when the number of pilots increases. However, a system with a large amount of pilot symbols is not of practical interest.

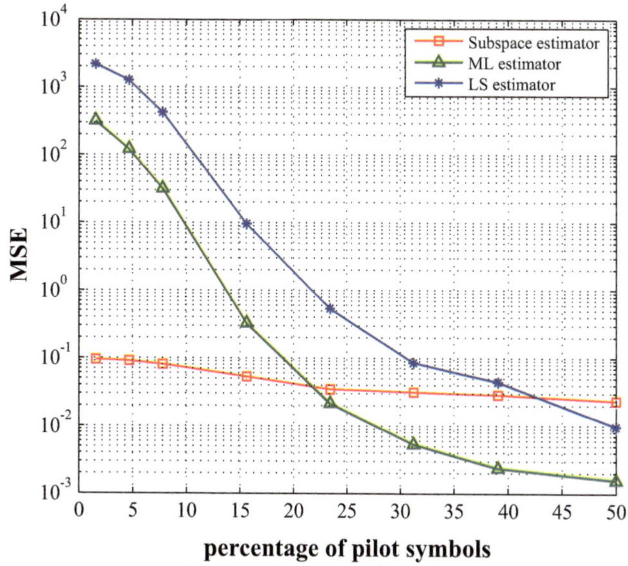

Fig. 5.4 Intended channel estimation MSE versus percentage of pilot symbols for SNR = 10 dB

In Figs. 5.5 and 5.6, we evaluate the impact of the number of observed OFDM symbols on the estimation performance. For the three algorithms, we consider the transmission scheme where the number of pilot symbols is set to one and the SNR is 10 dB. As the subspace algorithm is based on estimates of the second-order statistics of the received signal, its performance varies with the number of OFDM symbols. All three algorithms are able to estimate the SI channel with an error floor for the LS. The ML and subspace algorithms offer the similar performance. On the other hand, the LS estimator fails to recover the intended channel, for any number of OFDM symbols. This can be explained by the fact that the number of unknowns (intended channel coefficients) is larger than the number of pilot symbols. Hence, it is not possible to use this method when the number of pilot symbols is small. The ML estimator presents also poor estimation performance for the intended channel, while the subspace method is able to return a good channel estimate, with a better bandwidth efficiency compared to the other estimators, as soon as there are enough OFDM symbols to compute the covariance matrix.

Our primary motivation of this work is to develop an accurate channel estimator to cancel the SI signal. The performance of the SI-canceller is represented by its achieved output signal-to-residual-SI-and-noise power ratio (SINR) after SI-cancellation versus the input SNR. Ideally, if SI could be completely cancelled then the residual SI after cancellation is 0, and consequently the output SINR equals the input SNR as is shown by the dashed line "perfect cancellation" in Fig. 5.7. In other words, the "perfect cancellation" is considered as the ideal upper-bound for the SINR. As shown in Fig. 5.7, with three iterations, the proposed subspace-based

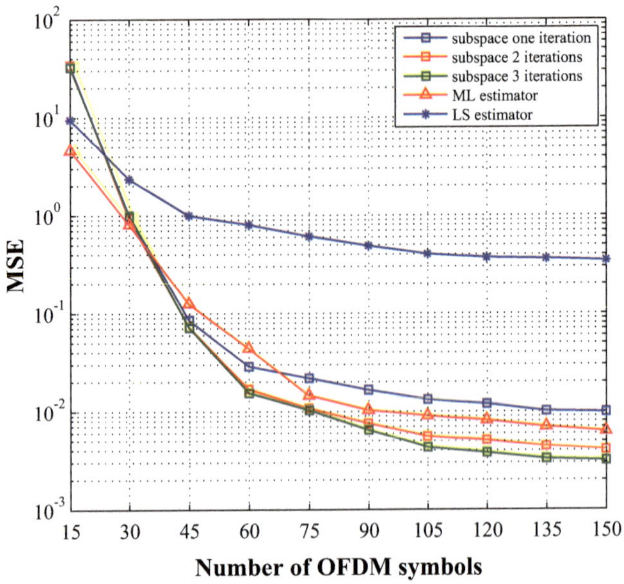

Fig. 5.5 SI channel estimation MSE versus number of OFDM symbols for SNR = 10 dB and one pilot symbol

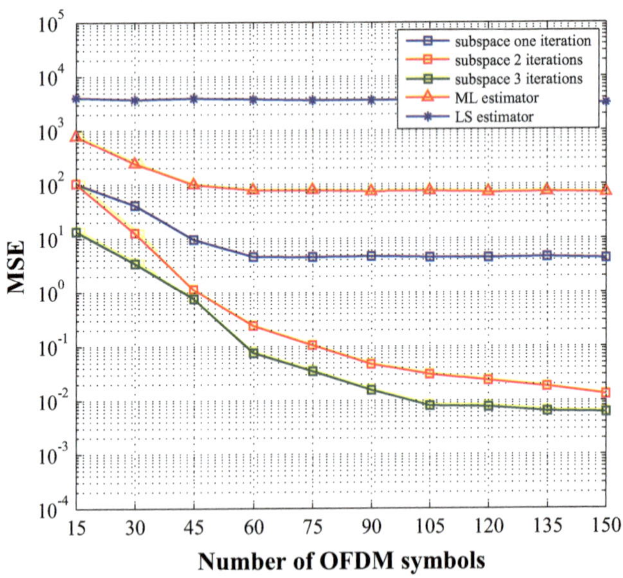

Fig. 5.6 Intended channel estimation MSE versus number of OFDM symbols for SNR = 10 dB and one pilot symbol

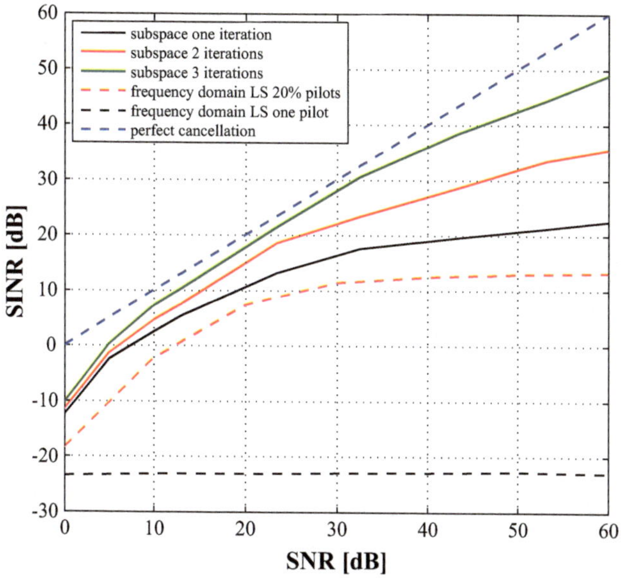

Fig. 5.7 Output SINR versus input SNR after SI-cancellation

SI-canceller can offer an output SINR very close to the upper-bound over a large SNR range. At low SNR, the large estimation error results in a larger residual SI after cancellation, which ultimately affects the output SINR.

We also investigate in Fig. 5.7 a frequency-domain method to estimate the different parameter using the pilot symbols on some subcarriers. We resort to the LS estimator to find the channel responses at the pilot subcarriers. Since the remaining subcarriers contain unknown symbols from the intended transceiver, the complete channel responses are obtained by linear interpolation of the estimated coefficients. Thus the frequency-domain approach uses only the portion of the signal containing pilots while the proposed approach exploits the whole received signal through the second-order statistics. As shown in Fig. 5.7, the performance of the frequency-domain approach highly depends on the number of pilots since the interpolation cannot model the variance of the channel in the frequency-domain. On the other hand, by exploiting the whole received signal through its second-order statistics, the proposed method offers good performance even with one pilot and still outperforms the frequency-domain approach (even with much larger number of pilots). Figure 5.8 plots the BER versus SNR curves of the two approaches. To improve the BER, the SINR should be kept as high as possible at the demodulator. To conclude, while the frequency-domain approach is more intuitive, it needs a large number of pilots and is outperformed by the proposed method.

We evaluate the performance of the system in the presence of phase noise by simulation. Figures 5.9 and 5.10 plot, respectively, the SINR and the BER versus the phase-noise 3 dB bandwidth $f_{3\,\mathrm{dB}}$ for SNR $=$ 20 dB, and common oscillator at the

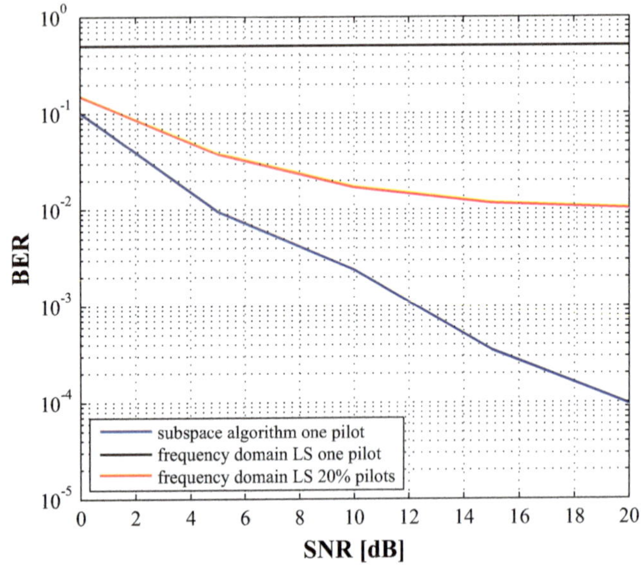

Fig. 5.8 BER versus SNR comparison of the proposed and the frequency-domain LS techniques

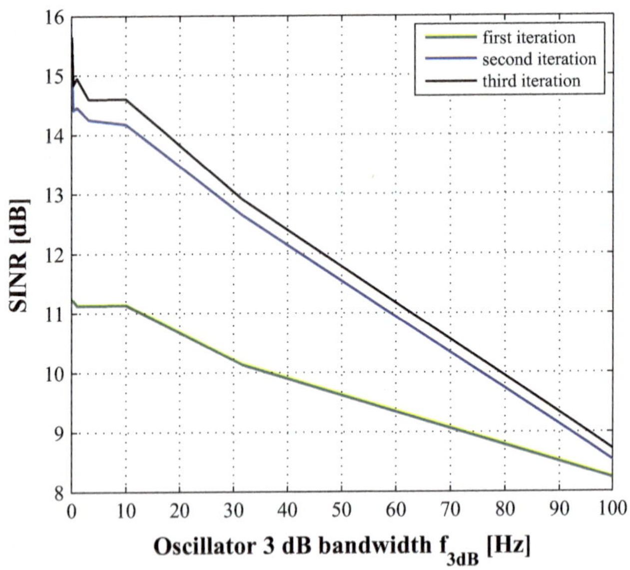

Fig. 5.9 SINR after SI-cancellation vs. $f_{3\,\mathrm{dB}}$

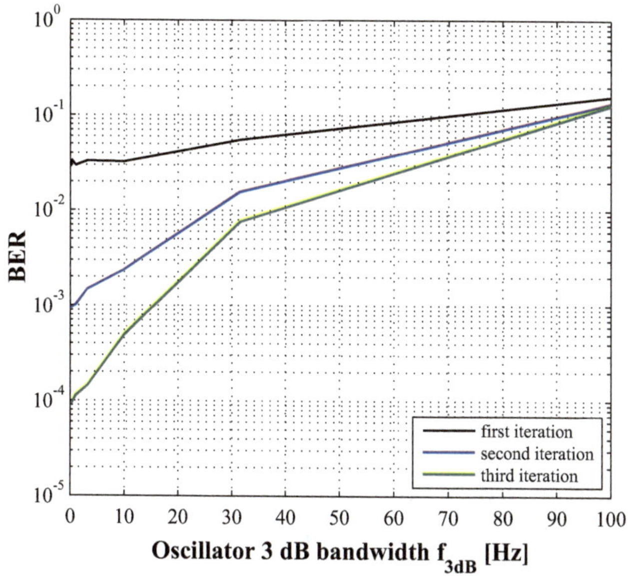

Fig. 5.10 BER vs. phase-noise $f_{3\,\mathrm{dB}}$

transmitter and the receiver. The residual SI depends on the quality of the oscillator represented by its $f_{3\,\mathrm{dB}}$. Higher $f_{3\,\mathrm{dB}}$ results in a fast varying process. From these figures, the proposed method still offers good cancellation performance, which is degraded as $f_{3\,\mathrm{dB}}$ increases.

The PA nonlinearity effects on the performance of the proposed algorithm are also investigated through simulations. Figure 5.11 plots the resulting SINR after cancellation versus the value of the PA IIP3 for SNR= 20 dB. For *perfect* cancellation, the resulting SINR after cancellation would be the SNR= 20 dB. A lower IIP3 indicates higher PA distortions (or poorer PA) and hence reduces the resulting SINR after cancellation. Figure 5.11 shows that as the IIP3 value increases, the cancellation performance is improved. However, for a sufficiently high IIP3 (e.g., 18 dBm or higher), the PA distortions are no longer dominant and the resulting SINR after cancellation is unchanged. This can be explained by the fact that, when developing the algorithm, the third-order component of the signal $x_{q,ip3}(n) = x_q^{IQ}(n)|x_q^{IQ}(n)|^2$ is approximated by $x_q(n)|x_q(n)|^2$ to simplify the algorithm. This approximation only affects the algorithm performance when the nonlinear coefficients are sufficiently high.

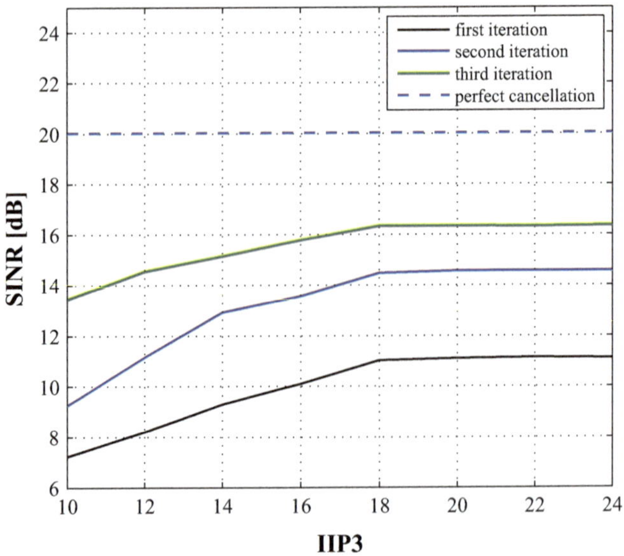

Fig. 5.11 SINR after SI-cancellation versus PA IIP3

5.4 Chapter Summary

In this chapter, a subspace-based estimation has been proposed to jointly estimate the SI channel, the intended channel and the transmitter impairments for MIMO full-duplex systems. By exploiting the covariance and pseudo-covariance matrices of the received signal, an effective way has been formulated to apply the subspace method for symmetric MIMO systems. The complete characterization of the second-order statistics of the received signal avoids the constraint on the number of transmit and receive antennas stated in Chap. 4. While the widely-linear formulation is appropriate for both situations, it is more convenient to select the direct subspace algorithm of proposed in Chap. 4 to reduce the size of the manipulated vectors and matrices.

References

1. W. H. Gerstacker, R. Schober, and A. Lampe, "Receivers with widely linear processing for frequency-selective channels," *IEEE Trans. on Comm.*, vol. 51, no. 9, pp. 1512–1523, 2003.
2. R. Schober, W. H. Gerstacker, and L.-J. Lampe, "Data-aided and blind stochastic gradient algorithms for widely linear MMSE MAI suppression for DS-CDMA," *IEEE Trans. Signal Process.*, vol. 52, no. 3, pp. 746–756, 2004.
3. P. Chevalier and A. Blin, "Widely linear MVDR beamformers for the reception of an unknown signal corrupted by noncircular interferences," *IEEE Trans. Signal Process.*, vol. 55, no. 11, pp. 5323–5336, 2007.

4. M. Kristensson, B. Ottersten, and D. Slock, "Blind subspace identification of a BPSK communication channel," in *Proc. ASILOMAR Signals, Syst., Comput.*, Pacific Frove, California, USA, Nov. 1996.
5. G. Gelli, L. Paura, and A. R. Ragozini, "Blind widely linear multiuser detection," *IEEE Comm. Lett.*, vol. 4, no. 6, pp. 187–189, 2000.
6. A. Masmoudi and T. Le-Ngoc, "Channel estimation and self-interference cancellation in full-duplex communication systems," *IEEE Trans. Vehicular Technology*, vol. 66, no. 1, pp. 321–334, Jan. 2017.
7. E. De Carvalho and D. T. Slock, "Cramer-Rao bounds for semi-blind, blind and training sequence based channel estimation," in *First IEEE Signal Processing Workshop on Signal Processing Advances in Wireless Communications*, Paris, France, Apr. 1997.
8. A. Masmoudi and T. Le-Ngoc, "Self-interference cancellation for full-duplex MIMO transceivers," in *Proc. IEEE Wireless Commun. and Netw. Conf.*, New Orleans, Louisiana, USA, Mar. 2015.
9. ——, "Subspace-based self-interference cancellation for full-duplex MIMO transceivers," *EURASIP Journal on Wireless Communications and Networking*, vol. 55, 2017. doi:10.1186/s13638-017-0839-x.
10. F. D. Neeser and J. L. Massey, "Proper complex random processes with applications to information theory," *IEEE Trans. Inf. Theory*, vol. 39, no. 4, pp. 1293–1302, 1993.
11. A. Masmoudi and T. Le-Ngoc, "A digital subspace-based self-interference cancellation in full-duplex MIMO transceivers," in *Proc. IEEE Int. Conf. Commun.*, London, UK, June 2015.
12. F. Gao, Y. Zeng, A. Nallanathan, and T.-S. Ng, "Robust subspace blind channel estimation for cyclic prefixed MIMO ODFM systems: Algorithm, identifiability and performance analysis," *IEEE J. Select. Areas in Comm.*, vol. 26, no. 2, pp. 378–388, 2008.
13. "1900 MHz high linearity amplifier." Application Note 1320, Avago Technologies.
14. "LTE; evolved universal terrestrial radio access (E-UTRA); user equipment (UE) radio transmission and reception (3GPP TS 36.101 version 11.2.0 release 11)." ETSI, Sophia Antipolis Cedex, France.
15. J. I. Choi, M. Jain, K. Srinivasan, P. Levis, and S. Katti, "Achieving single channel, full duplex wireless communication," in *Proc. ACM MobiCom*, Chicago, Illinois, USA, Sept. 2010.
16. A. Masmoudi and T. Le-Ngoc, "A maximum-likelihood channel estimator for self-interference cancellation in full-duplex systems," *IEEE Trans. Vehicular Technology*, vol. 65, no. 7, pp. 5122–5132, July 2016.

Chapter 6
Maximum Likelihood Self-Interference Cancellation

Abstract In this chapter, we propose a maximum likelihood (ML) approach to jointly estimate the self-interference and the intended channels. We perfectly exploit the known transmitted symbols from its own transceiver and both the known pilot and unknown data symbols from the other intended transceiver. The ML solution is obtained by maximizing the ML function under the assumption of Gaussian received symbols. A closed-form solution is first derived, and subsequently an iterative procedure is developed to further improve the estimation performance at moderate to high signal-to-noise ratio (SNR). We establish the initial condition to guarantee the convergence of the iterative algorithm to the ML solution. In the presence of considerable phase noise from the oscillators, a method to estimate this phase noise process is proposed which is, combined with the ML channel estimator, able to mitigate the effects of the phase noise. Illustrative results show that the proposed methods perform close to the Cramer-Rao bound (CRB) and offer good cancellation performance.

In Chaps. 4 and 5, we rely on the subspace approach to estimate the SI parameters for the baseband cancellation. This approach was first motivated by the need to incorporate the intended signal in the estimation process. Unlike the transmitted SI, the intended signal is not known beforehand. Therefore, the estimation process is based on the statistics of the received signal [1, 2]. In this chapter, we jointly estimate the SI channel, the intended channel and the transmitter nonlinearities, for the baseband cancellation stage, using the maximum-likelihood (ML) criterion when a set of subcarriers is reserved for pilots transmission. Since the received signal contains a mix of known and unknown data, the developed estimator exploits these known data and the second-order statistics of the unknown data from the intended transceiver towards the identification of the channels. The full use of the received signal reduces the number of needed pilot symbols compared to *training based* techniques. The received signal is approximated by a Gaussian process to formulate the likelihood function. Using some approximations, we derive a closed-form solution to maximize the likelihood function. A substantial improvement in estimation accuracy is obtained by iteratively estimating the second-order statistics of the unknown signal and the unknown coefficients.

T. Le-Ngoc, A. Masmoudi, *Full-Duplex Wireless Communications Systems*,
Wireless Networks, DOI 10.1007/978-3-319-57690-9_6

As stated in Chap. 3, the transmitter nonlinearities have to be reduced in the baseband cancellation stage. However, the phase noise from the local oscillators can also result in high residual SI [3, 4]. A shared-oscillator reduces the phase noise effects and improves the cancellation performance by 25 dB compared to two separate-oscillators for the up-conversion and down-conversion [5]. In this case, the difference between the phase noise at the transmit and receive chains depends on the delay that the SI signal experiences from the transmit chain to the receive chain. A frequency-domain method to compensate such phase noise is proposed in [6] and a time-domain phase noise estimation technique is developed in [7]. These methods consider the intended signal as an additive noise, which reduces the estimation accuracy. In this chapter, once an initial estimate of the channel coefficients is obtained, we propose a ML estimate of the phase noise affecting both the SI and intended signals, which avoids the drawback of considering the intended signal an additive noise.

The remaining of the chapter is organized as follows. In Sect. 6.1, we present the full-duplex communication system model under consideration. The analysis and development of the proposed ML channel estimation algorithm are presented in Sect. 6.2 and the procedure to estimate the phase noise process is detailed in Sect. 6.3. Section 6.4 provides illustrative simulation results and Sect. 6.5 presents the conclusion.

6.1 System Model

Consider that P_{pilot} subcarriers are dedicated to transmit pilot symbols. We define the index set of the subcarrier reserved for pilots by $\mathscr{P} = \{p_1, \ldots, p_{P_{\text{pilot}}}\}$, then the transmit signal $s_{q,t}(n)$ can be represented as the sum of the following two signals:

$$\begin{aligned} s^{p}_{q,t}(n) &= \frac{1}{\sqrt{N}} \sum_{i=1}^{P_{\text{pilot}}} S_{q,t}(p_i) e^{j2\pi p_i n/N}, \\ s^{d}_{q,t}(n) &= \frac{1}{\sqrt{N}} \sum_{k \notin \mathscr{P}} S_{q,t}(k) e^{j2\pi kn/N}, \end{aligned} \tag{6.1}$$

for $n = 0, \ldots, N-1$ where the first sequence $s^{p}_{q,t}(n)$ contains the pilot symbols $S_{q,t}(p_i)$, $p_i \in \mathscr{P}$, and the second sequence $s^{d}_{q,t}(n)$ contains the unknown transmit data symbols $S_{q,t}(k)$, $k \notin \mathscr{P}$, during the tth OFDM block. Using (6.1) and following the same model as in Chap. 4, the received signal in (4.29) becomes:

$$y_{r,t}(n) = \sum_{q=1}^{N_t} \sum_{l=0}^{L} h^{rsi}_{r,q}(l) x_{q,t}(n-l) + h^{rsi,iq}_{r,q}(l) x^{*}_{q,t}(n-l)$$

$$+\sum_{p=1}^{P} h_{r,q}^{rsi,ip,2p+1}(l)x_{q,t}(n-l)|x_{q,t}(n-l)|^{2p}$$
$$+h_{r,q}^{rsi,ip,2p+1,iq}(l)x_{q,t}^{*}(n-l)|x_{q,t}(n-l)|^{2p}$$
$$+h_{r,q}^{s}(l)s_{q,t}^{p}(n-l) + h_{r,q}^{s}(l)s_{q,t}^{d}(n-l) + w_{r,t}(n). \tag{6.2}$$

In (6.2), the equivalent global channel responses of the IQ image is:

$$h_{r,q}^{rsi,iq}(l) = g_{2,q}h_{r,q}^{rsi}(l), \tag{6.3}$$

and the global channel responses for the PA distortion and the combined PA and IQ distortions are given by, respectively:

$$\begin{aligned} h_{r,q}^{rsi,ip,2p+1}(l) &= \alpha_{2p+1,q}h_{r,q}^{rsi}(l), \\ h_{r,q}^{rsi,ip,2p+1,iq}(l) &= \alpha_{2p+1,q}g_{2,q}h_{r,q}^{rsi}(l). \end{aligned} \tag{6.4}$$

To allow a more articulate description of the problem, we define the set of $N \times (L+1)$ circulant matrices $\boldsymbol{X}_{q,cir,t}$, for $q = 1, \ldots, N_t$ in which the first row is $[x_{q,t}(0),\ x_{q,t}(N-1),\ x_{q,t}(N-2), \ldots,\ x_{q,t}(N-L)]$ and first column is given by $[x_{q,t}(0),\ x_{q,t}(1), \ldots,\ x_{q,t}(N-1)]$, and we define the $N \times N_t(L+1)$ matrix $\boldsymbol{X}_t = [\boldsymbol{X}_{1,cir,t},\ \boldsymbol{X}_{2,cir,t}, \ldots,\ \boldsymbol{X}_{N_t,cir,t}]$. The matrices $\boldsymbol{S}_t^p$ and $\boldsymbol{X}_{t,ip,2p+1}$ are defined in the same way as $\boldsymbol{X}_t$ but using the sequence $\{s_{q,t}^p(n)\}$ and $\{x_{q,t}(n)|x_{q,t}(n)|^{2p}\}$ instead of $\{x_{q,t}(n)\}$, respectively. We also gather the channel coefficients from all the transmit antennas to the rth receive antenna as:

$$\begin{aligned} \boldsymbol{h}_r^{rsi} &= \left[h_{r,1}^{rsi}(0), \ldots,\ h_{r,1}^{rsi}(L), \ldots,\ h_{r,N_t}^{rsi}(0), \ldots,\ h_{r,N_t}^{rsi}(L)\right]^T, \\ \boldsymbol{h}_r^{s} &= \left[h_{r,1}^{s}(0), \ldots,\ h_{r,1}^{s}(L), \ldots,\ h_{r,N_t}^{s}(0), \ldots,\ h_{r,N_t}^{s}(L)\right]^T, \\ \boldsymbol{H}_r^{s} &= \left[\boldsymbol{H}_{r,1}^{s},\ \boldsymbol{H}_{r,2}^{s}, \ldots,\ \boldsymbol{H}_{r,N_t}^{s}\right], \end{aligned} \tag{6.5}$$

where the $N \times N$ circulant matrix $\boldsymbol{H}_{r,q}^{s}$ is defined as:

$$\boldsymbol{H}_{r,q}^{s} = \begin{pmatrix} h_{r,q}^{s}(0) & 0 & \ldots & 0 & h_{r,q}^{s}(L) & \ldots & h_{r,q}^{s}(1) \\ h_{r,q}^{s}(1) & \ddots & & & & \ddots & \vdots \\ \vdots & & & & & & h_{r,q}^{s}(L) \\ h_{r,q}^{s}(L) & \ldots & h_{r,q}^{s}(0) & & & & 0 \\ \vdots & \ddots & & & \ddots & & \vdots \\ 0 & & h_{r,q}^{s}(L) & & \ldots & & h_{r,q}^{s}(0) \end{pmatrix}.$$

The vectors $\boldsymbol{h}_r^{rsi,iq}$, $\boldsymbol{h}_r^{rsi,ip,2p+1}$ and $\boldsymbol{h}_r^{rsi,ip,2p+1,iq}$ are defined in the same way as $\boldsymbol{h}_r^{rsi}$ in (6.5) using $h_{r,q}^{rsi,iq}(l)$, $h_{r,q}^{rsi,ip,2p+1}(l)$ and $h_{r,q}^{rsi,ip,2p+1,iq}(l)$, respectively. Using the previous notations, the received signal at antenna r can be reformulated in vector form as:

$$\boldsymbol{y}_{r,t} = \boldsymbol{X}_t\boldsymbol{h}_r^{rsi} + \boldsymbol{X}_t^*\boldsymbol{h}_r^{rsi,iq} + \sum_{p=1}^{P}\left(\boldsymbol{X}_{t,ip,2p+1}\boldsymbol{h}_r^{rsi,ip,2p+1} + \boldsymbol{X}_{t,ip,2p+1}^*\boldsymbol{h}_r^{rsi,ip,2p+1,iq}\right) + \boldsymbol{S}_t^p\boldsymbol{h}_r^s + \boldsymbol{H}_r^s\boldsymbol{s}_t^d + \boldsymbol{w}_{r,t}, \quad (6.6)$$

where $\boldsymbol{y}_{r,t} = [y_{r,t}(0), \ldots, y_{r,t}(N-1)]^T$ is the received $N \times 1$ vector after removing the cyclic prefix and $\boldsymbol{s}_t^d = \left[s_{q,t}^d(0), \ldots, s_{q,t}^d(N-1)\right]^T$. In the following, we limit our analysis to the third order PA nonlinearity. This is done to simplify the notation and to make it more illustrative. The generalization for any nonlinearity order can be derived from the following development by putting the vectors $h_r^{rsi,ip,2p+1,iq}$ on top of each other. By collecting the received vectors from the N_r receiving antennas in $\boldsymbol{y}_t = \left[\boldsymbol{y}_{1,t}^T, \ldots, \boldsymbol{y}_{N_r,t}^T\right]^T$, we can express (6.6) as:

$$\boldsymbol{y}_t = (\boldsymbol{I}_{N_r} \otimes \boldsymbol{X}_t)\boldsymbol{h}^{rsi} + (\boldsymbol{I}_{N_r} \otimes \boldsymbol{X}_t^*)\boldsymbol{h}^{rsi,iq} + (\boldsymbol{I}_{N_r} \otimes \boldsymbol{X}_{t,ip,3})\boldsymbol{h}^{rsi,ip,3} + (\boldsymbol{I}_{N_r} \otimes \boldsymbol{X}_{t,ip,3}^*)\boldsymbol{h}^{rsi,ip,3,iq} + (\boldsymbol{I}_{N_r} \otimes \boldsymbol{S}_t^p)\boldsymbol{h}^s + \boldsymbol{H}^s\boldsymbol{s}_t^d + \boldsymbol{w}_t, \quad (6.7)$$

where $\otimes$ refers to the Kronecker product between two matrices, $\boldsymbol{I}_{N_r}$ is the $N_r \times N_r$ identity matrix, the intended channel coefficients are collected as:

$$\boldsymbol{h}^s = \left[\boldsymbol{h}_1^{sT}, \boldsymbol{h}_2^{sT}, \ldots, \boldsymbol{h}_{N_r}^{sT}\right]^T,$$
$$\boldsymbol{H}^s = \left[\boldsymbol{H}_1^{sT}, \boldsymbol{H}_2^{sT}, \ldots, \boldsymbol{H}_{N_r}^{sT}\right]^T, \quad (6.8)$$

$\boldsymbol{h}^{rsi}$ is defined as:

$$\boldsymbol{h}^{rsi} = \left[\boldsymbol{h}_1^{rsi^T}, \boldsymbol{h}_2^{rsi^T}, \ldots, \boldsymbol{h}_{N_r}^{rsi^T}\right]^T, \quad (6.9)$$

and $\boldsymbol{h}^{rsi,iq}$, $\boldsymbol{h}^{rsi,ip,3}$ and $\boldsymbol{h}^{rsi,ip,3,iq}$ are defined in the same way. In the following, we assume that the noise and the transmitted signals are independent, and the signal and noise variances are α^2 and σ^2, respectively.

6.2 ML Estimator

To reduce the SI in (6.7), we need to estimate the residual SI channel $\boldsymbol{h}^{rsi}$ and the various equivalent channels from the transmitter impairments from the received signal $\boldsymbol{y}_t$. In this chapter, we propose a joint estimation of the SI and intended channels, exploiting both the known pilot symbols and the statistics of the unknown part of the received signal. The use of the known and unknown transmit data in the estimation process is commonly referred as semi-blind channel estimation [8, 9]. To that end, we introduce $\boldsymbol{h} = \left[\boldsymbol{h}^{rsi^T}, \boldsymbol{h}^{rsi,iq^T}, \boldsymbol{h}^{rsi,ip,3^T}, \boldsymbol{h}^{rsi,ip,3,iq^T}, \boldsymbol{h}^{s^T}\right]^T$ as the vector to be estimated and $\boldsymbol{D}_t = [\boldsymbol{I}_{N_r} \otimes \boldsymbol{X}_t, \boldsymbol{I}_{N_r} \otimes \boldsymbol{X}_t^*, \boldsymbol{I}_{N_r} \otimes \boldsymbol{X}_{t,ip,3}, \boldsymbol{I}_{N_r} \otimes \boldsymbol{X}_{t,ip,3}^*, \boldsymbol{I}_{N_r} \otimes \boldsymbol{S}_t^p]$ as the matrix gathering the symbols sent by the same transceiver and the known pilot symbols sent by the other intended transceiver. It follows that the received signal in (6.7) can be simply formulated as:

$$\boldsymbol{y}_t = \boldsymbol{D}_t \boldsymbol{h} + \boldsymbol{H}^s \boldsymbol{s}_t^d + \boldsymbol{w}_t. \tag{6.10}$$

For a Gaussian received data,[1] $\boldsymbol{y}_t$ is a Gaussian random vector with mean $\boldsymbol{D}_t\boldsymbol{h}$ and covariance matrix $\boldsymbol{R} = \alpha^2 \boldsymbol{H}^s \boldsymbol{H}^{sH} + \sigma^2 \boldsymbol{I}_{NN_r}$. A total of T OFDM symbols are used in the estimation process. Following the Gaussian model, the log-likelihood function is given by:

$$\mathscr{L}(\boldsymbol{h}) = -T \log |\boldsymbol{R}| - \sum_{t=1}^{T} (\boldsymbol{y}_t - \boldsymbol{D}_t\boldsymbol{h})^H \boldsymbol{R}^{-1} (\boldsymbol{y}_t - \boldsymbol{D}_t\boldsymbol{h}). \tag{6.11}$$

The ML estimate of $\boldsymbol{h}$ is obtained by maximizing the log-likelihood function $\mathscr{L}(\cdot)$. As the covariance matrix $\boldsymbol{R}$ depends on the unknown vector $\boldsymbol{h}^s$, maximizing the cost function with respect to $\boldsymbol{h}^s$ appears to be computationally intractable since it involves a $N_t N_r (L + 1)$-dimensional grid search. To overcome this complexity, we first ignore the relation between $\boldsymbol{R}$ and $\boldsymbol{h}^s$ and we maximize the log-likelihood function with respect to $\boldsymbol{R}$ and $\boldsymbol{h} = \left[\boldsymbol{h}^{rsi^T}, \boldsymbol{h}^{rsi,iq^T}, \boldsymbol{h}^{rsi,ip,3^T}, \boldsymbol{h}^{rsi,ip,3,iq^T}, \boldsymbol{h}^{s^T}\right]^T$. This separability is exploited to solve the problem in a low-complexity manner. In the following, we derive a closed-form solution and an iterative method to estimate the channels.

6.2.1 Closed-Form Solution

By considering separable variables $\boldsymbol{h}$ and $\boldsymbol{R}$, the conditional approach to maximize the log-likelihood function can be used. In the conditional approach, the covariance matrix is modeled as deterministic and unknown. Therefore, the matrix $\boldsymbol{R}$ is substituted by the solution $\boldsymbol{R}_{ML}(\boldsymbol{h})$ that maximizes (6.11) for a fixed $\boldsymbol{h}$. Hence, maximizing (6.11) with respect to $\boldsymbol{R}$ leads to [11, 12]:

[1] The Gaussian assumption is well justified for OFDM transmit signal [10].

$$\boldsymbol{R}_{ML}(\boldsymbol{h}) = \frac{1}{T}\sum_{t=1}^{T}(\boldsymbol{y}_t - \boldsymbol{D}_t\boldsymbol{h})(\boldsymbol{y}_t - \boldsymbol{D}_t\boldsymbol{h})^H. \tag{6.12}$$

Substituting $\boldsymbol{R}$ by $\boldsymbol{R}_{ML}(\boldsymbol{h})$ in (6.11), we get the so-called compressed likelihood function [13, 14]:

$$\mathscr{L}_c(\boldsymbol{h}) = -T\log\left|\frac{1}{T}\sum_{t=1}^{T}(\boldsymbol{y}_t - \boldsymbol{D}_t\boldsymbol{h})(\boldsymbol{y}_t - \boldsymbol{D}_t\boldsymbol{h})^H\right|, \tag{6.13}$$

where the constant terms irrelevant to the maximization have been discarded. It follows that the ML channel estimate is given by:

$$\boldsymbol{h}_{ML} = \arg\max_{\boldsymbol{h}} \mathscr{L}_c(\boldsymbol{h}). \tag{6.14}$$

In order to find a closed-form solution of (6.14), we first compute the LS estimate of the channel [15]:

$$\boldsymbol{h}_{LS} = \left(\sum_{t=1}^{T}\boldsymbol{D}_t^H\boldsymbol{D}_t\right)^{-1}\sum_{t=1}^{T}\boldsymbol{D}_t^H\boldsymbol{y}_t. \tag{6.15}$$

Then, we define $\boldsymbol{d}_t = \boldsymbol{y}_t - \boldsymbol{D}_t\boldsymbol{h}_{LS}$ and $\widetilde{\boldsymbol{R}} = 1/T\sum_{t=1}^{T}\boldsymbol{d}_t\boldsymbol{d}_t^H$. Following theses notations, the compressed likelihood function in (6.13) can be rewritten as [1]:

$$\mathscr{L}_c(\boldsymbol{h}) = -T\log\left|\widetilde{\boldsymbol{R}} + \frac{1}{T}\sum_{t=1}^{T}\boldsymbol{D}_t(\boldsymbol{h}-\boldsymbol{h}_{LS})(\boldsymbol{h}-\boldsymbol{h}_{LS})^H\boldsymbol{D}_t^H - \boldsymbol{D}_t(\boldsymbol{h}-\boldsymbol{h}_{LS})\boldsymbol{d}_t^H - \boldsymbol{d}_t(\boldsymbol{h}-\boldsymbol{h}_{LS})^H\boldsymbol{D}_t^H\right|. \tag{6.16}$$

Let define $\boldsymbol{\xi} = \boldsymbol{h} - \boldsymbol{h}_{LS}$. As the block number T increases, the LS estimate $\boldsymbol{h}_{LS}$ approaches the ML estimate $\boldsymbol{h}_{ML}$. Therefore, the difference $\boldsymbol{\xi}_{ML} = \boldsymbol{h}_{ML} - \boldsymbol{h}_{LS}$ between the two estimates becomes small. Using the fact that for any matrix $\boldsymbol{M}$ satisfying $||\boldsymbol{M}||_F \ll 1$, we have $|\boldsymbol{I}+\boldsymbol{M}| \approx 1 + \text{trace}\{\boldsymbol{M}\}$, the log-likelihood function in (6.16) is rearranged to obtain:

$$\mathscr{L}_c(\boldsymbol{h}) = -T\log|\widetilde{\boldsymbol{R}}| - T\left(1 + \frac{1}{T}\text{trace}\left\{\widetilde{\boldsymbol{R}}^{-1}\sum_{t=1}^{T}\boldsymbol{D}_t\boldsymbol{\xi}\boldsymbol{\xi}^H\boldsymbol{D}_t^H - \boldsymbol{D}_t\boldsymbol{\xi}\boldsymbol{d}_t^H - \boldsymbol{d}_t\boldsymbol{\xi}^H\boldsymbol{D}_t^H\right\}\right), \tag{6.17}$$

where we substitute $\boldsymbol{h} - \boldsymbol{h}_{LS}$ by $\boldsymbol{\xi}$. Using the commutative property of the trace, the maximization of $\mathscr{L}_c(\boldsymbol{h})$ is equivalent to:

$$\boldsymbol{\xi}_{ML} = \arg\max_{\boldsymbol{\xi}} \sum_{t=1}^{T} \boldsymbol{\xi}^H \boldsymbol{D}_t^H \widetilde{\boldsymbol{R}}^{-1} \boldsymbol{D}_t \boldsymbol{\xi} - \boldsymbol{d}_t^H \widetilde{\boldsymbol{R}}^{-1} \boldsymbol{D}_t \boldsymbol{\xi} - \boldsymbol{\xi}^H \boldsymbol{D}_t^H \widetilde{\boldsymbol{R}}^{-1} \boldsymbol{d}_t. \tag{6.18}$$

By setting the first derivative with respect to $\boldsymbol{\xi}$ to zero, the solution to (6.18) is given by:

$$\boldsymbol{\xi}_{ML} = \left(\sum_{t=1}^{T} \boldsymbol{D}_t^H \widetilde{\boldsymbol{R}}^{-1} \boldsymbol{D}_t \right)^{-1} \sum_{t=1}^{T} \boldsymbol{D}_t^H \widetilde{\boldsymbol{R}}^{-1} \boldsymbol{d}_t. \tag{6.19}$$

Using $\boldsymbol{d}_t = \boldsymbol{y}_t - \boldsymbol{D}_t \boldsymbol{h}_{LS}$ and $\boldsymbol{h}_{ML} = \boldsymbol{h}_{LS} + \boldsymbol{\xi}_{ML}$, a closed-form expression of the ML channel estimate is given by:

$$\boldsymbol{h}_{ML} = \left(\sum_{t=1}^{T} \boldsymbol{D}_t^H \widetilde{\boldsymbol{R}}^{-1} \boldsymbol{D}_t \right)^{-1} \sum_{t=1}^{T} \boldsymbol{D}_t^H \widetilde{\boldsymbol{R}}^{-1} \boldsymbol{y}_t. \tag{6.20}$$

For large data record, we show in Appendix 1 that $\lim_{T\to\infty} \widetilde{\boldsymbol{R}} = \boldsymbol{R}$. Considering the expression of $\boldsymbol{y}_t$ given in (6.10), the estimator in (6.20) is expanded as:

$$\boldsymbol{h}_{ML} = \boldsymbol{h} + \left(\sum_{t=1}^{T} \boldsymbol{D}_t^H \widetilde{\boldsymbol{R}}^{-1} \boldsymbol{D}_t \right)^{-1} \sum_{t=1}^{T} \boldsymbol{D}_t^H \widetilde{\boldsymbol{R}}^{-1} \left(\boldsymbol{H}^s \boldsymbol{s}_t^d + \boldsymbol{w}_t \right), \tag{6.21}$$

from where we can verify that the estimator is asymptomatically unbiased.

The ML estimate obtained in (6.20) is different from the LS estimate because of the weighting matrix $\widetilde{\boldsymbol{R}}^{-1}$. Actually, the ML and LS estimates are equivalent in the presence of white Gaussian noise. In our case, the effective noise is composed of the thermal noise and unknown transmit signal, which is not a white noise. In an alternative interpretation, the ML solution can be viewed as a weighted LS solution where the covariance matrix of the noise affecting $\boldsymbol{y}_t$ is exactly $\boldsymbol{R}$ and we use $\widetilde{\boldsymbol{R}}$ as an estimate of $\boldsymbol{R}$.

6.2.2 *Iterative ML Estimator*

The closed-form solution in (6.20) depends on $\widetilde{\boldsymbol{R}}$, which is an estimate of the covariance matrix $\boldsymbol{R}$. Therefore, a better estimate of $\boldsymbol{R}$ results in a better estimate of the channel vector $\boldsymbol{h}$. On the other hand, the matrix $\boldsymbol{R}$ depends on the unknown intended channel coefficients $\boldsymbol{h}^s$ that we want to estimate. Assuming again the separability of the log-likelihood function in $\boldsymbol{h}$ and $\boldsymbol{R}$, a common approach in this situation is to resort to an iterative procedure. If the channel vector is given, the covariance matrix $\boldsymbol{R}$ that maximizes the log-likelihood function given $\boldsymbol{h}$ is:

$$\boldsymbol{R}_{ML}(\boldsymbol{h}) = \frac{1}{T}\sum_{t=1}^{T}(\boldsymbol{y}_t - \boldsymbol{D}_t\boldsymbol{h})(\boldsymbol{y}_t - \boldsymbol{D}_t\boldsymbol{h})^H. \quad (6.22)$$

Conversely, if $\boldsymbol{R}$ is available, the solution to the problem $\arg\max_{\boldsymbol{h}} \mathscr{L}(\boldsymbol{h}, \boldsymbol{R})$ can be computed as:

$$\boldsymbol{h}_{ML}(\boldsymbol{R}) = \left(\sum_{t=1}^{T}\boldsymbol{D}_t^H\boldsymbol{R}^{-1}\boldsymbol{D}_t\right)^{-1}\sum_{t=1}^{T}\boldsymbol{D}_t^H\boldsymbol{R}^{-1}\boldsymbol{y}_t. \quad (6.23)$$

The proposed approach iterates between (6.22) and (6.23). At the ith iteration, the estimate $\boldsymbol{R}_{i-1}$ obtained at iteration $i-1$ is used to find $\boldsymbol{h}$ as $\boldsymbol{h}_i = \boldsymbol{h}_{ML}(\boldsymbol{R}_{i-1})$. Then, the estimate of $\boldsymbol{R}$ is updated at iteration i as $\boldsymbol{R}_i = \boldsymbol{R}_{ML}(\boldsymbol{h}_i)$. The algorithm is stopped when there is no significant difference between two consecutive estimates. Like most of iterative algorithms, initialization is a critical issue for convergence. In our case, setting $\boldsymbol{R}_0 = \boldsymbol{I}$ appears to be a reasonable starting point. At the first iteration, we obtain the LS estimate given in (6.15). As we iterate, the matrix $\boldsymbol{R}_i$ acts as a weighting matrix to improve the estimated channel.

The proof of convergence to the global maximum of the log-likelihood function may not be straightforward because the function at hand is not verified to be convex. However, using the closed-form expression obtained in the previous section, it is possible to prove the convergence to a stationary point close to the ML solution. In fact, when initializing the algorithm with $\boldsymbol{R}_0 = \boldsymbol{I}$, the iterative algorithm gives us, in the second iteration, the same channel estimate given in the closed-form solution in (6.20). That is, after two iterations, the algorithm operates close to the ML solution. Following the arguments in [16], we have:

$$\begin{aligned}\mathscr{L}(\boldsymbol{h}_i, \boldsymbol{R}_i) &= \max_{\boldsymbol{R}} \mathscr{L}(\boldsymbol{h}_i, \boldsymbol{R}) \\ &\geq \mathscr{L}(\boldsymbol{h}_i, \boldsymbol{R}_{i-1}) \\ &= \max_{\boldsymbol{h}} \mathscr{L}(\boldsymbol{h}, \boldsymbol{R}_{i-1}) \\ &\geq \mathscr{L}(\boldsymbol{h}_{i-1}, \boldsymbol{R}_{i-1}).\end{aligned} \quad (6.24)$$

Therefore, the log-likelihood function is increased after each iteration, and for a good initialization, the algorithm converges to a stationary point close to the ML solution. Simulation results presented in Sect. 6.4 confirm that, when initializing the algorithm with $\boldsymbol{R}_0 = \boldsymbol{I}$, the iterative algorithm converges after four to five iterations. The main complexity of the LS estimator comes from the computation of the inverse of $\left(\sum_{t=1}^{T}\boldsymbol{D}_t^H\boldsymbol{D}_t\right)$ while the proposed ML algorithm involves an additional matrix inversion of the $NN_r \times NN_r$ matrix $\boldsymbol{R}_i$ at each iteration compared to the LS estimator.

Table 6.1 Comparison of the baseband estimation algorithms

	Linear subspace algorithm	Widely-linear subspace algorithm	ML algorithm
Main complexity	– Eigen-decomposition: $\mathcal{O}(MN_r)^3 + \mathcal{O}(N_r(L+1)p)^3$ – Ambiguity term: $\mathcal{O}(2N_t)^2 + \mathcal{O}(6N_t^2)^2$	– Eigen-decomposition: $\mathcal{O}(2N_rM)^3 + \mathcal{O}(4N_tN_r(L+1))^3$ – Ambiguity term: $\mathcal{O}((N+p+1)N_t)^3 + \mathcal{O}(4N_tN_r)^3$	Matrix inversion: $\mathcal{O}(N_{\text{iter}}(5(L+1)+NN_r)^3)$
Application cases	– Number of receive antennas N_r should be double number of transmit antennas N_t – No need for pilot symbols	– Number of receive antennas N_r can be equal to the number of transmit antennas N_t – No need for pilot symbols	– No restriction on number of antennas – Need some pilot symbols

Table 6.1 provides a complexity comparison of the estimation algorithms proposed for the baseband cancellation stage, with the corresponding application since every algorithm has its own requirements in term of number of antennas and pilot symbols.

6.3 Phase Noise Suppression

While the knowledge of the SI channel and the transmitter impairments are essential to reconstruct the received SI signal, some other RF components can affect the SI and, it is desired to consider their effects when cancelling the received SI. Phase noise, introduced by both transmitter and receiver oscillators, has been considered as one of the main limiting factors in SI-cancellation. Considering the presence of phase noise in the received signal, (6.2) can be rewritten as:

$$y_{r,t}(n) = \left(\sum_{q=1}^{N_t}\sum_{l=0}^{L} h^{rsi}_{r,q}(l)\left(x_{q,t}(n-l)+x^{imp}_{q,t}(n-l)\right)e^{j\phi^i_{n-l,q}}\right.$$
$$\left. + h^s_{r,q}(l)s_{q,t}(n-l)e^{j\phi^s_{n-l,q}}\right)e^{-j\phi_{n,r}} + w_{r,t}(n), \tag{6.25}$$

where $\phi^i_{n,q}$ is the phase noise of its own qth oscillator of the transmitter side affecting the nth SI sample, $\phi^s_{n,q}$ is the phase noise of the qth intended transmitter oscillator, affecting the nth intended received sample, $\phi_{n,r}$ is the oscillator phase noise at the rth receive antenna affecting both the SI and intended signals, and $x^{imp}_{q,t}(n)$ collects the transmitter impairments from the IQ mixer and the PA. For generality, we use different notations for the phase noise process from different antennas, allowing us to apply the proposed method for independent oscillators at the different antennas or a common shared-oscillator. The phase noise processes in (6.25) changes from one OFDM symbol to another and should be indexed by time t, but this notation is ignored for clarity while we keep in mind that the phase noises change from one ODFM symbol to another.

Since the transmitted symbols multiplied by different phase noise realisations are further convolved by the multipath channel impulse response, the received sample n is affected by $L+1$ different realizations of phase noise. However, the phase noise, due to oscillator imperfection, is typically a very narrowband process and, hence, changes slowly over time. As a result, the difference in phases during these $L+1$ consecutive symbols can be assumed to be negligible in order to simplify the development of the algorithm, i.e., $\phi^i_{n-l,q} = \phi^i_{n,q}$ and $\phi^s_{n-l,q} = \phi^s_{n,q}$, for $l = 0, \ldots, L$. Therefore, the received signal in (6.25) becomes:

$$y_{r,t}(n) = \sum_{q=1}^{N_t}\sum_{l=0}^{L} h^{rsi}_{r,q}(l)\Big(x_{q,t}(n-l)+x^{imp}_{q,t}(n-l)\Big)\, e^{j\phi^i_{n,r,q}} + h^s_{r,q}(l)\left(s^p_{q,t}(n-l)+s^d_{q,t}(n-l)\right) e^{j\phi^s_{n,r,q}} + w_{r,t}(n), \tag{6.26}$$

where $\phi^i_{n,q,r} = \phi^i_{n,q} - \phi_{n,r}$ and $\phi^s_{n,q,r} = \phi^s_{n,q} - \phi_{n,r}$ are the combined transmit and receive phase noise processes affecting the SI and the intended signals, respectively. It is noteworthy to mention that we adopt this assumption only during the development of the algorithms and simulations are performed using the actual phase noise process. Denoting $\widetilde{x}_{r,q,t}(n) = \sum_{l=0}^{L} h^{rsi}_{r,q}(l)(x_{q,t}(n-l) + x^{imp}_{q,t}(n-l))$ the received SI and the transmitter impairments in the absence of phase noise, also $\widetilde{s}^p_{r,q,t}(n) = \sum_{l=0}^{L} h^s_{r,q}(l)s^p_{q,t}(n-l)$ and $\widetilde{s}^d_{r,q,t}(n) = \sum_{l=0}^{L} h^s_{r,q}(l)s^d_{q,t}(n-l)$, the N received samples of the tth OFDM block can be expressed as:

$$\boldsymbol{y}_{r,t} = \sum_{q=1}^{N_t} \mathrm{diag}\{\widetilde{x}_{r,q,t}(n)\}\boldsymbol{\Phi}^i_{r,q} + \left(\mathrm{diag}\{\widetilde{s}^p_{r,q,t}(n)\} + \mathrm{diag}\{\widetilde{s}^d_{r,q,t}(n)\}\right)\boldsymbol{\Phi}^s_{r,q} + \boldsymbol{w}_{r,t}, \tag{6.27}$$

where diag $\{\widetilde{x}_{r,q,t}(n)\}$ is a diagonal matrix with diagonal elements $\{\widetilde{x}_{r,q,t}(0), \ldots, \widetilde{x}_{r,q,t}(N-1)\}$ and:

$$\boldsymbol{\Phi}^i_{r,q} = \left[e^{j\phi^i_{0,q,r}}, e^{j\phi^i_{1,r,q}}, \ldots, e^{j\phi^i_{N-1,r,q}}\right]^T,$$

$$\boldsymbol{\Phi}^s_{r,q} = \left[e^{j\phi^s_{0,r,q}}, e^{j\phi^s_{1,q,r}}, \ldots, e^{j\phi^s_{N-1,q,r}}\right]^T. \tag{6.28}$$

Assuming an estimate of the channel coefficients is available, from the proposed estimator in Sect. 6.2, the joint estimation of the phase noise vectors $\boldsymbol{\Phi}^i_{r,q}$ and $\boldsymbol{\Phi}^s_{r,q}$, for $q = 1, \ldots, N_t$, involves recovering $2NN_t$ parameters from N equations. This is an underdetermined problem and may have many different solutions. Thus, exploiting the fact that the phase noise is slowly varying over time, we consider that it remains constant over l_p consecutive samples, which divides the number of unknowns by l_p. Let $\overline{\boldsymbol{\Phi}}^i_{r,q}$ and $\overline{\boldsymbol{\Phi}}^s_{r,q}$ denote the reduced version of $\boldsymbol{\Phi}^i_{r,q}$ and $\boldsymbol{\Phi}^s_{r,q}$ whose elements are defined as $\overline{\boldsymbol{\Phi}}^i_{r,q}(n) = \boldsymbol{\Phi}^i_{r,q}(nl_p)$ and $\overline{\boldsymbol{\Phi}}^s_{r,q}(n) = \boldsymbol{\Phi}^s_{r,q}(nl_p)$, respectively, and let:

$$\widetilde{\boldsymbol{X}}_{r,q,t} = \begin{pmatrix} \widetilde{x}_{r,q,t}(0) & 0 & \dots & 0 \\ \vdots & \vdots & & \\ \widetilde{x}_{r,q,t}(l_p-1) & 0 & & \vdots \\ 0 & \widetilde{x}_{r,q,t}(l_p) & & \\ \vdots & \vdots & \ddots & \\ 0 & \widetilde{x}_{r,q,t}(2l_p-1) & & 0 \\ & & & \widetilde{x}_{r,q,t}(N-l_p) \\ \vdots & & \ddots & \vdots \\ 0 & & & \widetilde{x}_{r,q,t}(N-1) \end{pmatrix},$$

$\widetilde{\boldsymbol{S}}^p_{r,q,t}$ and $\widetilde{\boldsymbol{S}}^d_{r,q,t}$ are defined in the same way as $\widetilde{\boldsymbol{X}}_{r,q,t}$ using the pilot part $\widetilde{s}^p_{q,t}(n)$ and the unknown part $\widetilde{s}^d_{q,t}(n)$ of the intended signal. Denoting $\widetilde{\boldsymbol{X}}_{r,t} = \left[\widetilde{\boldsymbol{X}}_{r,1,t}, \dots, \widetilde{\boldsymbol{X}}_{r,N_t,t}\right]$ and $\widetilde{\boldsymbol{S}}^p_{r,t} = \left[\widetilde{\boldsymbol{S}}^p_{r,1,t}, \dots, \widetilde{\boldsymbol{S}}^p_{r,N_t,t}\right]$, the received vector in (6.27) can be approximated by:

$$\begin{aligned} \boldsymbol{y}_{r,t} &\approx \sum_{q=1}^{N_t} \widetilde{\boldsymbol{X}}_{r,q,t}\overline{\boldsymbol{\Phi}}^i_{r,q} + \widetilde{\boldsymbol{S}}^p_{r,q,t}\overline{\boldsymbol{\Phi}}^s_{r,q} + \widetilde{\boldsymbol{S}}^d_{r,q,t}\overline{\boldsymbol{\Phi}}^s_{r,q} + \boldsymbol{w}_{r,t}, \\ &= \widetilde{\boldsymbol{X}}_{r,t}\overline{\boldsymbol{\Phi}}^i_r + \widetilde{\boldsymbol{S}}^p_{r,t}\overline{\boldsymbol{\Phi}}^s_r + \widetilde{\boldsymbol{S}}^d_{r,t}\overline{\boldsymbol{\Phi}}^s_r + \boldsymbol{w}_{r,t}, \end{aligned} \tag{6.29}$$

with $\overline{\boldsymbol{\Phi}}^i_r = \left[\overline{\boldsymbol{\Phi}}^{iT}_{r,1}, \dots, \overline{\boldsymbol{\Phi}}^{iT}_{r,N_t}\right]^T$ and $\overline{\boldsymbol{\Phi}}^s_r = \left[\overline{\boldsymbol{\Phi}}^{sT}_{r,1}, \dots, \overline{\boldsymbol{\Phi}}^{sT}_{r,N_t}\right]^T$. Finally, by collecting the phase noise processes as:

$$\begin{aligned} \overline{\boldsymbol{\Phi}}^i &= \left[\overline{\boldsymbol{\Phi}}^{iT}_1, \overline{\boldsymbol{\Phi}}^{iT}_2 \dots, \overline{\boldsymbol{\Phi}}^{iT}_{N_r}\right]^T, \\ \overline{\boldsymbol{\Phi}}^s &= \left[\overline{\boldsymbol{\Phi}}^{sT}_1, \overline{\boldsymbol{\Phi}}^{sT}_2 \dots, \overline{\boldsymbol{\Phi}}^{sT}_{N_r}\right]^T, \end{aligned} \tag{6.30}$$

and defining the block diagonal matrices $\widetilde{\boldsymbol{X}}_t$, $\widetilde{\boldsymbol{S}}^p_t$ and $\widetilde{\boldsymbol{S}}^d_t$ with block diagonal elements $\widetilde{\boldsymbol{X}}_{r,t}$, $\widetilde{\boldsymbol{S}}^p_{r,t}$ and $\widetilde{\boldsymbol{S}}^d_{r,t}$, for $r = 1, \dots, N_r$, respectively, the received vector $\boldsymbol{y}_t = \left[\boldsymbol{y}^T_{1,t}, \dots, \boldsymbol{y}^T_{N_r,t}\right]^T$ over the N_r antennas can be written in the following compact form:

$$\boldsymbol{y}_t = \widetilde{\boldsymbol{X}}_t\overline{\boldsymbol{\Phi}}^i + \widetilde{\boldsymbol{S}}^p_t\overline{\boldsymbol{\Phi}}^s + \widetilde{\boldsymbol{S}}^d_t\overline{\boldsymbol{\Phi}}^s + \boldsymbol{w}_t. \tag{6.31}$$

Similar to Sect. 6.2, we gather the parameters to be estimated in one vector $\boldsymbol{\Phi} = \left[\overline{\boldsymbol{\Phi}}^{iT}, \overline{\boldsymbol{\Phi}}^{sT}\right]^T$ and the known transmitted signals in one matrix as $\widetilde{\boldsymbol{D}}_t = \left[\widetilde{\boldsymbol{X}}_t, \widetilde{\boldsymbol{S}}_t^p\right]$. Thus, by adopting the Gaussian model to the received vector $\boldsymbol{y}_t$, the log-likelihood function to estimate the phase noise, knowing the channel coefficients, is expressed as:

$$\mathscr{L}_{pn}\left(\overline{\boldsymbol{\Phi}}^i, \overline{\boldsymbol{\Phi}}^s\right) = -\log\left|\boldsymbol{R}_{pn}\right| - \left(\boldsymbol{y}_t - \widetilde{\boldsymbol{D}}_t\boldsymbol{\Phi}\right)^H \boldsymbol{R}_{pn}^{-1}\left(\boldsymbol{y}_t - \widetilde{\boldsymbol{D}}_t\boldsymbol{\Phi}\right), \tag{6.32}$$

where $\boldsymbol{R}_{pn}$ is the covariance matrix of $\boldsymbol{y}_t$ given by $\boldsymbol{R}_{pn} = \mathbb{E}\left\{\widetilde{\boldsymbol{S}}_t^d\overline{\boldsymbol{\Phi}}^s\overline{\boldsymbol{\Phi}}^{sH}\widetilde{\boldsymbol{S}}_t^{dH}\right\} + \sigma^2\boldsymbol{I}_{NN_r}$. It is noteworthy to mention that maximizing $\mathscr{L}_{pn}(\cdot,\cdot)$ with respect to $\overline{\boldsymbol{\Phi}}^i$ and $\overline{\boldsymbol{\Phi}}^s$ leads to the ML estimate of the phase noise processes. Compared to the problem of channel estimation in Sect. 6.2, the covariance matrix $\boldsymbol{R}_{pn}$ depends on the statistical properties of the phase noise process and not on its actual realization of the process. Therefore, the problem is reduced to minimizing $(\boldsymbol{y}_t - \widetilde{\boldsymbol{D}}_t\boldsymbol{\Phi})^H\boldsymbol{R}_{pn}^{-1}(\boldsymbol{y}_t - \widetilde{\boldsymbol{D}}_t\boldsymbol{\Phi})$ with respect to $\boldsymbol{\Phi}$. By setting the first derivative to zero, it can be shown that the ML estimate of the phase noises is given by:

$$\boldsymbol{\Phi}_{ML} = \left(\widetilde{\boldsymbol{D}}_t^H\boldsymbol{R}_{pn}^{-1}\widetilde{\boldsymbol{D}}_t\right)^{-1}\widetilde{\boldsymbol{D}}_t^H\boldsymbol{R}_{pn}^{-1}\boldsymbol{y}_t. \tag{6.33}$$

The closed form expression of $\boldsymbol{R}_{pn}$ depends on the oscillator type. In Appendix 2, the expression of $\boldsymbol{R}_{pn}$ is given for a phase-locked loop (PLL)-based oscillator and a free-running oscillator.

The main complexity of the phase noise estimation procedure comes from the inversion of the $2NN_tN_r/l_p \times 2NN_tN_r/l_p$ matrix $\widetilde{\boldsymbol{D}}_t^H\boldsymbol{R}_{pn}^{-1}\widetilde{\boldsymbol{D}}_t$. Note that the phase noise estimator needs also the inverse of the covariance matrix $\boldsymbol{R}_{pn}^{-1}$, which is computed one time only, since it depends on the characteristics of the oscillators only and not on the transmitted signal.

The matrix $\widetilde{\boldsymbol{D}}_t^H$ depends on the SI channel, the transmitter impairments and intended channel. An iterative technique is used to jointly estimate the channel coefficients and the phase noise required to suppress the SI signal. First, an initial estimate of the channel is obtained using the proposed ML algorithm by ignoring the presence of the phase noise. Then, the estimated channels are used to obtain an estimate of the phase noise vector $\boldsymbol{\Phi}$ from (6.33). Next, we use the estimate of the phase noise to shift the transmitted SI signal and intended signal and estimate the channel coefficients again from the shifted reference signal. We iterate this procedure until the algorithm converges.

As any iterative method, the convergence to the actual solution should be discussed. The proposed method may converge to a stationary point which is different from the actual channel coefficients and the phase noise processes. For example, if $h_{r,q}^{rsi}$ and $e^{j\phi_{n,r,p}^i}$ are solutions to our estimation problem, then $h_{r,q}^{rsi}e^{j\beta}$ and $e^{j(\phi_{n,r,p}^i - \beta)}$ would also be solutions. Actually, the phase noise process at time n is written as $\phi_{n,r,q}^i = \phi_{n-1,r,q}^i + \delta_{n,r,q}^i$ where $\delta_{n,r,q}^i$ is the innovation process at time n.

Thus the phase noise process $\phi^i_{n,r,q}$ can be expressed as the sum of a constant term $\phi^i_{0,r,q}$ and a variable term $\Delta^i_{n,r,q} = \sum_{k=1}^{n} \delta^i_{k,r,q}$. When combined with the propagation channel, it is not possible to separate the phase of the channel coefficients and $\phi^i_{0,r,q}$. Therefore, the channel estimation algorithm returns an estimate of $h^{rsi}_{r,q} e^{j\phi^i_{0,r,q}}$ and $h^s_{r,q} e^{j\phi^s_{0,r,q}}$ while leaving the variable part of the phase noise $\Delta^i_{n,r,q}$ and $\Delta^s_{n,r,q}$ to be estimated during the second part of the procedure. The iterative algorithm is more likely to converge to the points $(h^{rsi}_{r,q} e^{j\phi^i_{0,r,q}}, \Delta^i_{n,r,q})$ than the actual points $(h^{rsi}_{r,q}, \phi^i_{n,r,q})$. A more detailed study on convergence is presented in Appendix 3.

6.4 Simulation Results

In this section, we provide some simulation results to illustrate the performance of the proposed algorithm in terms of the estimation error and the SI-cancellation capability in different scenarios. The wireless channels are represented as a frequency selective Rayleigh fading channel with 9 equal-variance resolvable paths ($L = 8$). The SIR_{in} at the input of the RF cancellation stage is assumed to be -50 dB. A first estimate of the SI channel is obtained during an initial half-duplex period as:

$$\widehat{\boldsymbol{h}}^{RF} = \left((\boldsymbol{I}_{N_r} \otimes \boldsymbol{X}_t)^H (\boldsymbol{I}_{N_r} \otimes \boldsymbol{X}_t)\right)^{-1} (\boldsymbol{I}_{N_r} \otimes \boldsymbol{X}_t)^H \boldsymbol{y}_t, \tag{6.34}$$

and the input to the proposed algorithm is the output of the RF cancellation stage. The data are drawn from 4-QAM constellation then passed through an OFDM modulator. Unless otherwise specified, the number of observed OFDM blocks is set to $T = 60$. The pilot symbols are inserted periodically in some subcarriers before the OFDM modulator.

6.4.1 *Performance in the Absence of Phase Noise*

We first evaluate the performance of the proposed channel estimator in the absence of phase noise. Figures 6.1 and 6.2 depict the MSE of the proposed method when estimating the residual SI $\boldsymbol{h}^{rsi}$ and the intended channel $\boldsymbol{h}^s$, respectively. The pilot symbols represent 20% of the total transmitted data. To properly assess the performance of the proposed channel estimator, the MSE of the algorithm is compared with the Cramér-Rao bound (CRB), as a benchmark for the performance evaluation of the estimator. The expression of the CRB is given in Appendix 4. We also compare the ML estimator with the LS estimator. The iterative ML algorithm is initialized with $\boldsymbol{R}_0 = \boldsymbol{I}$ in the absence of any prior information about the intended channel. As mentioned in the previous section, the LS estimate is obtained in the first iteration of the proposed algorithm. From the simulation results, the closed-

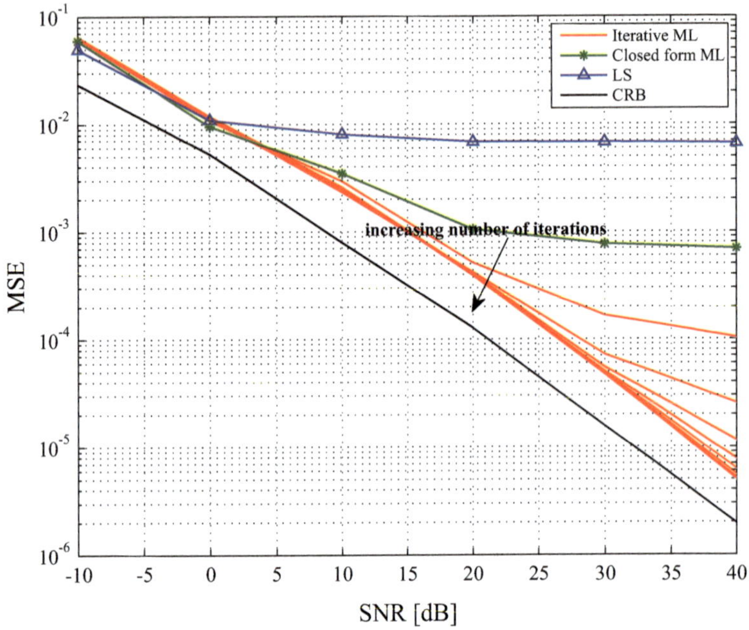

Fig. 6.1 SI channel estimation MSE vs. SNR with $N_t = N_r = 2$

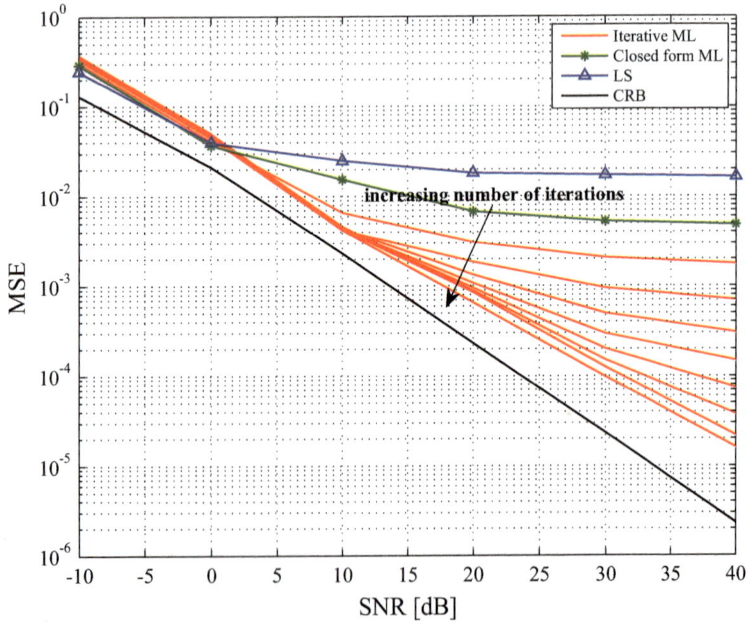

Fig. 6.2 Intended channel estimation MSE vs. SNR with $N_t = N_r = 2$

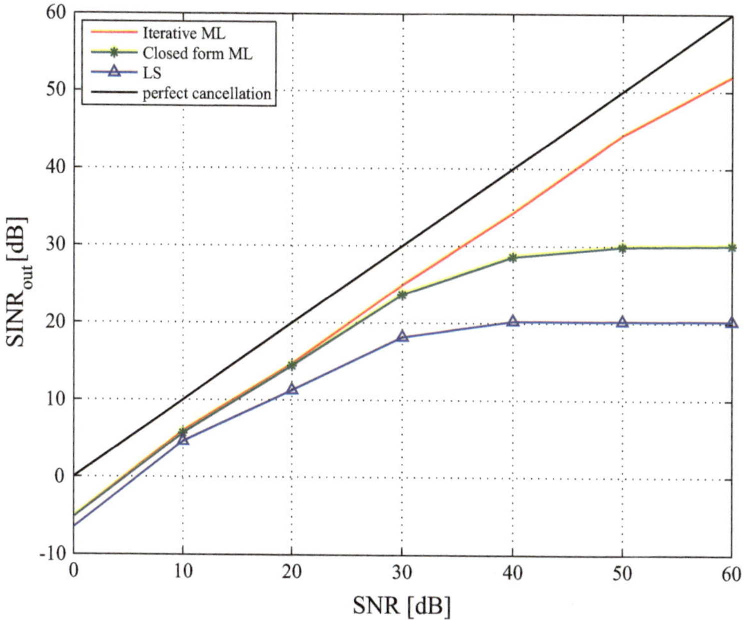

Fig. 6.3 SINR$_{out}$ after SI-cancellation vs. SNR

form ML and the LS perform close to the CRB for moderate SNR. However, the LS saturates at high SNR. This saturation is due to the presence of the unknown received signal from the intended user, which acts as noise floor as the SNR increases. Whereas, at low SNR, the thermal noise is dominant as compared to the unknown transmitted signal and the estimation performance is mostly affected by the thermal noise. The closed-form ML also presents a noise floor at high SNR because of the different approximations adopted to obtain the closed-form expression. As we iterate, this saturation is reduced since we have a better estimate of the covariance matrix $\boldsymbol{R}$. At low SNR, convergence is obtained after three or four iterations while more iterations are needed at high SNR.

Figure 6.3 represents the relation between the input SNR and output SINR (SINR$_{out}$) after SI-cancellation. It can be seen that the proposed iterative algorithm outperforms the closed-form ML solution at moderate and high SNR. Actually, the MSE saturation of the closed-form ML is reflected in SINR$_{out}$ since the cancellation performance is directly related to the accuracy of the estimated SI channel. Figure 6.4 shows the amount of baseband cancellation α_{BB}, defined as the SI power at the output of the RF stage divided by the remaining SI power after the baseband cancellation. Both iterative and closed-form solutions are compared to the LS estimator. Clearly, the proposed algorithm outperforms the LS estimator. Note that the performance of the LS estimator saturates at high interference power as a consequence of the channel estimation MSE saturation at high SNR shown in the previous simulations.

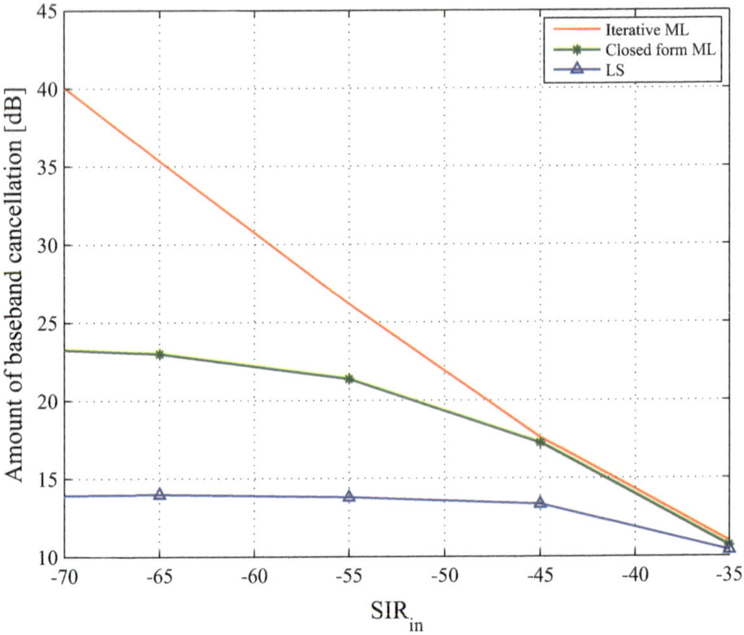

Fig. 6.4 Amount of baseband cancellation vs. SIR_{in}

The estimation of the intended channel needs some pilot symbols from the other transceiver. In Fig. 6.5, we evaluate the performance of the proposed algorithms when varying the number of pilots with SNR= 20 dB and with seven iterations for the iterative ML algorithm. The MSE for SI channel estimation (in solid lines) is less affected by the number of pilots than the intended channel estimation (in dashed lines). For a very small number of pilots, the SI channel is estimated by using the known *self-signal* and the statistics of the intended signal. For more pilots, the intended channel can be estimated and a more accurate estimate of the statistics of the intended signal can be obtained.

To investigate the effect of the RF cancellation on the performance of the proposed scheme, we investigate the relation between the two following parameters: (1) the RF cancellation gain α_{RF} defined as the ratio between the SI input and output powers of the RF cancellation stage, and (2) the baseband cancellation gain α_{BB}. The results plotted in Fig. 6.6 show that, the more α_{RF} increases, the more α_{BB} decreases. In fact, as α_{RF} increases, the amount of SI left for the baseband cancellation stage is reduced, leaving not much interference to be cancelled in the next stage.

Increasing number of transmitting antennas N_t and receive antennas N_r results in more channel coefficients between the different antennas, and thus, more parameters to estimate from a larger number of observations coming from the receive antennas. Thus, as the number of antennas increases, the size of the matrices involved in the closed-form and iterative solutions increases too. In fact, the matrix $\widetilde{\boldsymbol{R}}$ involved in

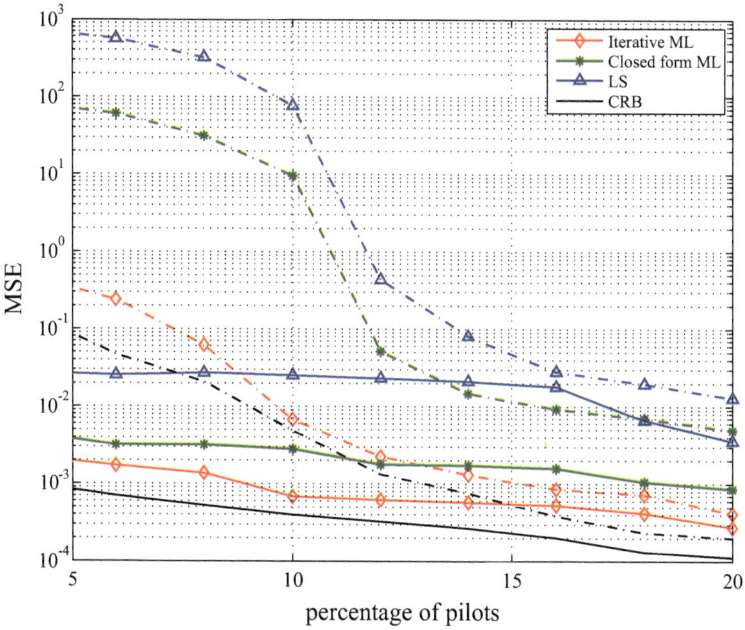

Fig. 6.5 SI channel (in *solid lines*) and intended channel (in *dashed lines*) estimation MSE vs. percentage of pilots

the closed-form solution has a size of $NN_r \times NN_r$, and its complexity increases with the number of iterations. On the other hand, one may intuitively think that, as the number of parameters increases, convergence would require more iterations. Table 6.2 summarizes the number of iterations required for convergence to estimate the SI and intended channels for different values of N_t and N_r (taking $N_t = N_r$) obtained from extensive simulations.

Figure 6.7 illustrates the ratio of the achievable rate of full-duplex to that of half-duplex versus the SNR using a SISO system. The rate of the full-duplex transmission is given by $R_{\mathrm{FD}} = 2\log(1 + \mathrm{SINR})$, where the SINR is obtained after the baseband cancellation using the widely-linear subspace algorithm of the iterative ML algorithm. Also, the rate of the half-duplex transmission is given by $R_{\mathrm{HD}} = 2\log(1 + \mathrm{SNR})$. At first, for systems operating at low SNR, full-duplex does not give much advantage over half-duplex. As the operating SNR increases, the full-duplex-to-half-duplex rate ratio increases to obtain more than 80% increase in spectral efficiency for certain values of SNR.

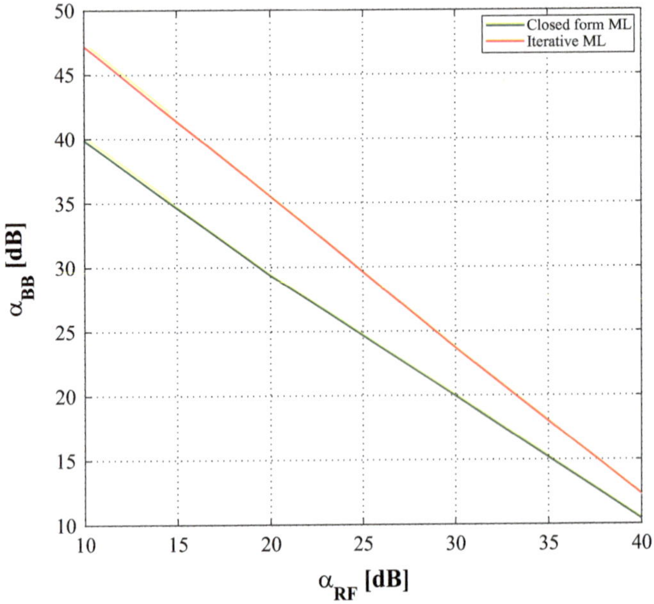

Fig. 6.6 Relationship between the cancellation gains of RF and baseband cancellation stages

Table 6.2 Number of iterations vs. number of antennas for SNR= 20 dB

Number of antennas	1	2	3	4	5
Number of iterations	4	6	7	9	10

6.4.2 Performance in the Presence of Phase Noise

In this subsection, we evaluate the performance of the proposed algorithm in the presence of phase noise. The estimate of the SI channel used in the RF cancellation stage is still obtained during the initial half-duplex period in the presence of all noise components. Figures 6.8 and 6.9 show the SINR_{out} for different values of SNR in the presence of phase noise with independent and shared PLL-based oscillator at the transmitters and receivers, respectively. The quality of the oscillator is often measured by its 3 dB bandwidth $f_{3\,\text{dB}}$. Higher 3 dB bandwidth level results in a fast varying process, making its estimation more difficult. In these figures, we set $f_{3\,\text{dB}} =$ 100 Hz. It is observed that the presence of phase noise reduces the SINR_{out} after cancellation, and the shared-oscillator case offers higher SINR_{out} than the separate-oscillators case. Actually, the effects of phase noise can be reduced by using the same common oscillator in the up-conversion and down-conversion of the same transceiver. In this case, the difference between the phase noises in the transmitter and the receiver depends on the delay that the SI experiences from the transmitter to the receiver. While the proposed method improves the cancellation performance by estimating the phase noise and mitigating its effects, a noise floor appears at high

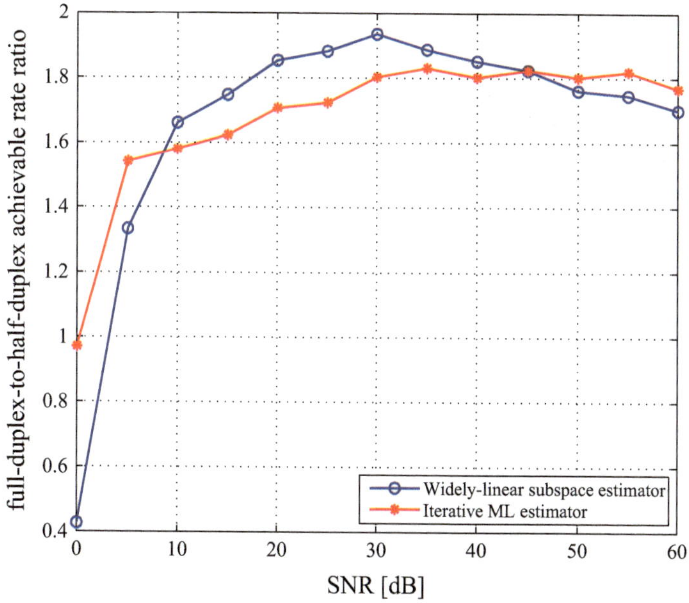

Fig. 6.7 Full-duplex-to-half-duplex achievable rate ratio vs. SNR

SNR. This noise floor comes from the approximation of the same phase noise over a set of samples used when developing the phase noise estimation algorithm. We also mention that a shared-oscillator is possible in practice when the transmit and the receive chains are located at the same transceiver.

The residual SI depends on the quality of the oscillators $f_{3\,\mathrm{dB}}$. The resulting SINR_{out} as a function of $f_{3\,\mathrm{dB}}$ is given in Fig. 6.10 for the common and separate-oscillator cases when the SNR is fixed at 35 dB. In the common oscillator case, the proposed algorithm can keep the SINR_{out} constant for $f_{3\,\mathrm{dB}}$ lower than 300 Hz. In the case with separate-oscillators, one can see that the phase noise makes the SINR_{out} increases starting from $f_{3\,\mathrm{dB}} = 100$ Hz.

6.4.3 Simulink Platform

A full-duplex wireless communications link simulator is implemented with Matlab and Simulink, using the SimRF library to include more realistic responses on the RF components. The following results support the extended simulations presented in Chaps. 5 and 6. We consider here an OFDM-SISO system with the general system level parameters shown in Table 6.3. Figure 6.11 represents a general view of the developed platform.

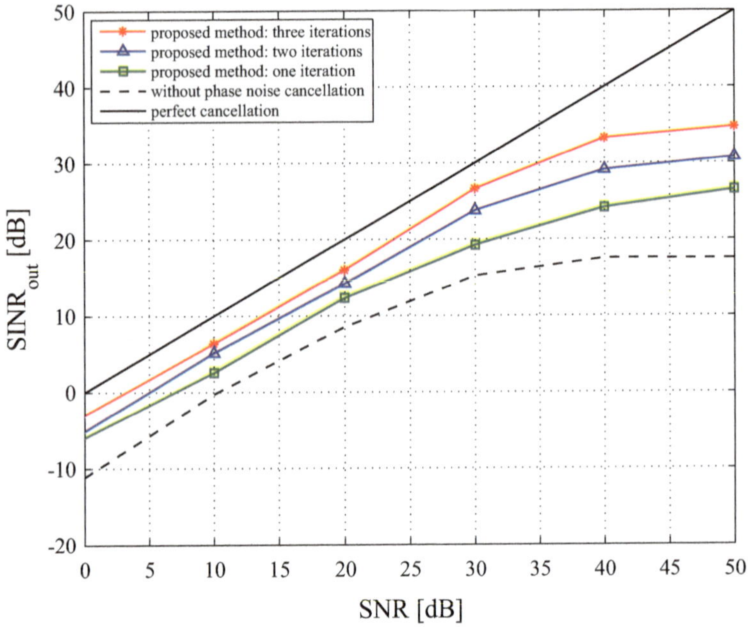

Fig. 6.8 SINR_{out} after SI-cancellation vs. SNR in the presence of phase noise from a shared-oscillator for $f_{3\,\text{dB}} = 100\,\text{Hz}$

Figures 6.12 and 6.13 show the spectrum plots of the SI at different points of the receiver. Figure 6.12 illustrates the case where only linear SI-cancellation is performed in the baseband and Fig. 6.13 illustrate the case where both linear and nonlinear cancellation are performed in the baseband cancellation stage. For comparison, we also plot the spectrum of the intended signal. The spectral regrowth due to transmitter impairments makes the SI higher than the intended signal if the transmitter impairments are not reduced in the baseband. Table 6.4 summarises the obtained BER from the Simulator platform when using the ML and the widely-linear subspace algorithms in the baseband. Table 6.4 indicates the BER results obtained from the Simulink platform and in Chap. 5 are in close agreement.

6.5 Chapter Summary

In this chapter, ML channel estimation in full-duplex MIMO transceivers has been investigated. A closed-form expression was obtained to jointly estimate the residual SI channel, transmitter impairments and the intended channel. An iterative procedure was also proposed to avoid the performance saturation of the closed-form solution as high SNR. The iterative procedure incorporates the statistics of the

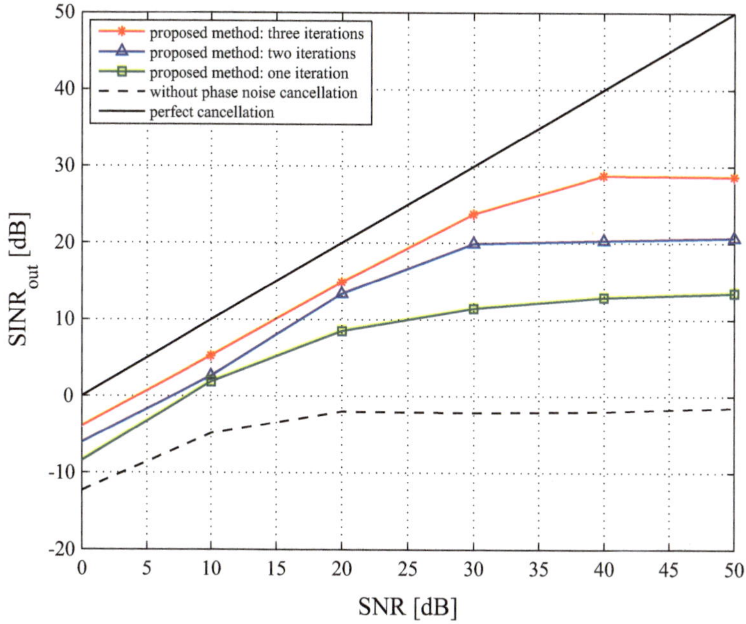

Fig. 6.9 SINR$_{out}$ after SI-cancellation vs. SNR in the presence of phase noise from separate-oscillators for $f_{3\,\mathrm{dB}} = 100\,\mathrm{Hz}$

unknown received signal to improve the estimation performance. In the presence of significant phase noise, a method that exploits the previous channel estimate was proposed to mitigate the effects of the phase noise.

Appendix 1: Proof of $\lim_{T\to\infty}\widetilde{\boldsymbol{R}} = \boldsymbol{R}$

In this appendix, we want to prove that $\lim_{T\to\infty}\widetilde{\boldsymbol{R}} = \lim_{T\to\infty} 1/T\boldsymbol{d}_t\boldsymbol{d}_t^H = \boldsymbol{R}$. First, $\boldsymbol{d}_t = \boldsymbol{y}_t - \boldsymbol{D}_t\boldsymbol{h}_{LS}$ can be written as:

$$\begin{aligned}
\boldsymbol{d}_t &= \boldsymbol{y}_t - \boldsymbol{D}_t\left(\sum_{n=1}^{T}\boldsymbol{D}_n^H\boldsymbol{D}_n\right)^{-1}\sum_{n=1}^{T}\boldsymbol{D}_n^H\boldsymbol{y}_n \\
&= \boldsymbol{D}_t\boldsymbol{h} + \boldsymbol{H}^s\boldsymbol{s}_t^d + \boldsymbol{w}_t - \boldsymbol{D}_t\left(\boldsymbol{h} + \left(\sum_{n=1}^{T}\boldsymbol{D}_n^H\boldsymbol{D}_n\right)^{-1}\sum_{n=1}^{T}\boldsymbol{D}_n^H\left(\boldsymbol{H}^s\boldsymbol{s}_n^d + \boldsymbol{w}_n\right)\right) \\
&= \boldsymbol{H}^s\boldsymbol{s}_t^d + \boldsymbol{w}_t - \boldsymbol{D}_t\left(\sum_{n=1}^{T}\boldsymbol{D}_n^H\boldsymbol{D}_n\right)^{-1}\left(\sum_{n=1}^{T}\boldsymbol{D}_n^H\left(\boldsymbol{H}^s\boldsymbol{s}_n^d + \boldsymbol{w}_n\right)\right).
\end{aligned} \tag{6.35}$$

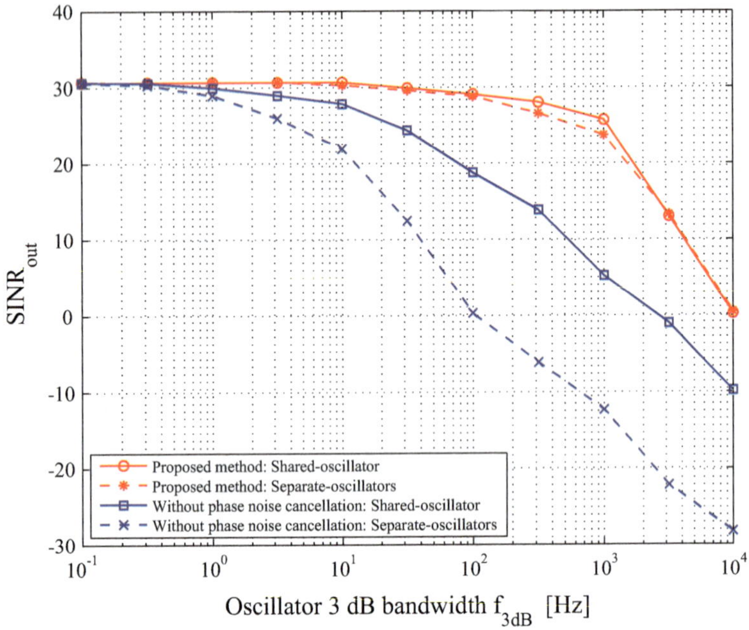

Fig. 6.10 SINR_{out} after SI-cancellation vs. 3 dB bandwidth of phase noise $f_{3\,\text{dB}}$

Table 6.3 System level parameters using in Simulink

Parameter	Value
Bandwidth	20 MHz
Carrier frequency	2.08 GHz
Sampling time	4×10^{-9} s
PA gain	30 dB
PA IIP2 and IIP3	45 and 20 dBm
ADC bits	12 dB
LNA gain	25 dB
LNA IIP2 and IIP3	43 dBm and -8 dBm
Constellation	4-QAM

Thus we obtain:

$$\boldsymbol{d}_t\boldsymbol{d}_t^H = \left(\boldsymbol{H}^s\boldsymbol{s}_t^d + \boldsymbol{w}_t\right)\left(\boldsymbol{H}^s\boldsymbol{s}_t^d + \boldsymbol{w}_t\right)^H - \boldsymbol{D}_t\left(\sum_{n=1}^{T}\boldsymbol{D}_n^H\boldsymbol{D}_n\right)^{-1}\left(\sum_{n=1}^{T}\boldsymbol{D}_n^H\left(\boldsymbol{H}^s\boldsymbol{s}_n^d + \boldsymbol{w}_n\right)\right)\left(\boldsymbol{H}^s\boldsymbol{s}_t^d + \boldsymbol{w}_t\right)^H$$
$$- \left(\boldsymbol{H}^s\boldsymbol{s}_t^d + \boldsymbol{w}_t\right)\left(\sum_{n=1}^{T}\left(\boldsymbol{H}^s\boldsymbol{s}_n^d + \boldsymbol{w}_n\right)^H\boldsymbol{D}_n\right)\left(\sum_{n=1}^{T}\boldsymbol{D}_n^H\boldsymbol{D}_n\right)^{-1}\boldsymbol{D}_t^H$$

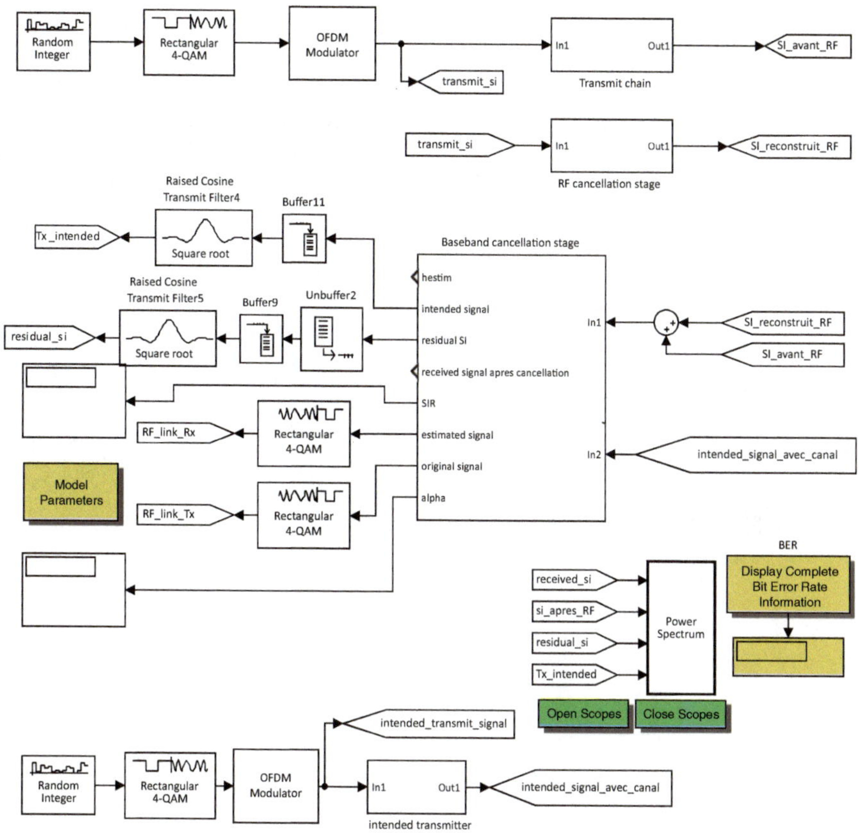

Fig. 6.11 High level block diagram of the Simulink platform

$$+\boldsymbol{D}_t\left(\sum_{n=1}^{T}\boldsymbol{D}_n^H\boldsymbol{D}_n\right)^{-1}\left(\sum_{n=1}^{T}\boldsymbol{D}_n^H\left(\boldsymbol{H}^s\boldsymbol{s}_n^d+\boldsymbol{w}_n\right)\right)\left(\sum_{n=1}^{T}\left(\boldsymbol{H}^s\boldsymbol{s}_n^d+\boldsymbol{w}_n\right)^H\boldsymbol{D}_n\right)\left(\sum_{n=1}^{T}\boldsymbol{D}_n^H\boldsymbol{D}_n\right)^{-1}\boldsymbol{D}_t^H. \tag{6.36}$$

By definition, we have $\widetilde{\boldsymbol{R}} = 1/T\sum_{t=1}^{T}\boldsymbol{d}_t\boldsymbol{d}_t^H$. Since:

$$\lim_{T\to\infty} 1/T\sum_{t=1}^{T}\left(\boldsymbol{H}^s\boldsymbol{s}_t^d+\boldsymbol{w}_t\right)\left(\boldsymbol{H}^s\boldsymbol{s}_t^d+\boldsymbol{w}_t\right)^H=\boldsymbol{R}, \tag{6.37}$$

the first term in $\widetilde{\boldsymbol{R}}$ converges to the covariance matrix of $\boldsymbol{H}^s\boldsymbol{s}_t^d+\boldsymbol{w}_t$. The limit of the second is:

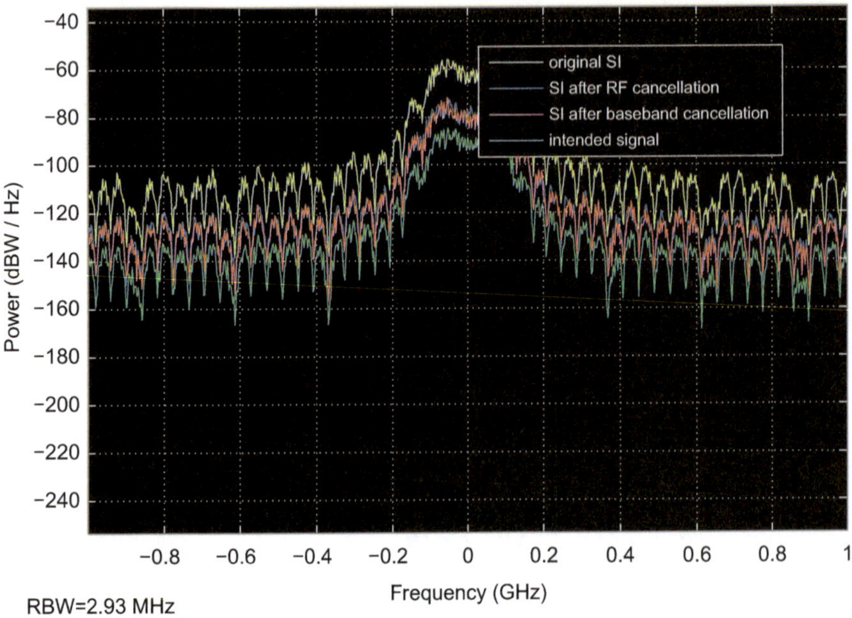

Fig. 6.12 SI spectrum without nonlinear cancellation

$$
\begin{aligned}
&\lim_{T\to\infty}\frac{1}{T}\sum_{t=1}^{T}\boldsymbol{D}_t\left(\sum_{n=1}^{T}\boldsymbol{D}_n^H\boldsymbol{D}_n\right)^{-1}\left(\sum_{n=1}^{T}\boldsymbol{D}_n^H\left(\boldsymbol{H}^s\boldsymbol{s}_n^d+\boldsymbol{w}_n\right)\right)\left(\boldsymbol{H}^s\boldsymbol{s}_t^d+\boldsymbol{w}_t\right)^H \\
&\quad=\lim_{T\to\infty}\frac{1}{T}\sum_{t=1}^{T}\boldsymbol{D}_t\left(\sum_{n=1}^{T}\boldsymbol{D}_n^H\boldsymbol{D}_n\right)^{-1}\left(\boldsymbol{D}_t^H\left(\boldsymbol{H}^s\boldsymbol{s}_t^d+\boldsymbol{w}_t\right)\right)\left(\boldsymbol{H}^s\boldsymbol{s}_t^d+\boldsymbol{w}_t\right)^H \\
&\quad=\mathbb{E}\left\{\boldsymbol{D}_t\lim_{T\to\infty}\left(\sum_{t=1}^{T}\boldsymbol{D}_t^H\boldsymbol{D}_t\right)^{-1}\boldsymbol{D}_t^H\right\}\boldsymbol{R} \\
&\quad=\boldsymbol{0},
\end{aligned}
\tag{6.38}
$$

where $\lim_{T\to\infty}\left(\sum_{t=1}^{T}\boldsymbol{D}_t^H\boldsymbol{D}_t\right)^{-1}=\boldsymbol{0}$. Following the same development, the other two terms converge also to $\boldsymbol{0}$. It results that $\lim_{T\to\infty}\widetilde{\boldsymbol{R}}=\boldsymbol{R}$.

Table 6.4 BER vs. SNR

SNR [dB]	BER using the widely-linear subspace algorithm	BER using the ML algorithm
0	0.095	0.1102
5	1.4×10^{-2}	1.8×10^{-2}
6	6.3×10^{-3}	6.1×10^{-3}
7	4.6×10^{-3}	4.8×10^{-3}
10	8.1×10^{-4}	1.12×10^{-3}

Appendix 2: Covariance Matrix $\widetilde{\boldsymbol{R}}_{pn}$

In this appendix, we give the expression of the covariance matrix $\widetilde{\boldsymbol{R}}_{pn}$ used to estimate the phase noise in (6.33). This is done in the cases of using PLL or free-running oscillators. First, for separate-oscillators in both the transmitter and receiver, the covariance of $e^{j\phi^s_{n,r,q}}$ and $e^{j\phi^s_{n',r',q'}}$ is given by:

$$\mathbb{E}\left\{e^{j\phi^s_{n,r,q}}e^{-j\phi^s_{n',r',q'}}\right\} = \mathbb{E}\left\{e^{j(\phi_{n,r}-\phi_{n',r'}+\phi^s_{n',q'}-\phi^s_{n,q})}\right\}$$

$$= \begin{cases} \mathbb{E}\left\{e^{j(\phi_{n,r}-\phi_{n',r})}\right\}\mathbb{E}\left\{e^{j(\phi^s_{n',q}-\phi^s_{n,q})}\right\}, & \text{for } q = q',\ r = r', \\ \mathbb{E}\left\{e^{j(\phi^s_{n',q}-\phi^s_{n,q})}\right\}e^{-\sigma^r_{pn}}, & \text{for } q = q',\ r \neq r', \\ e^{-\sigma^t_{pn}}\mathbb{E}\left\{e^{j(\phi_{n,r}-\phi_{n',r})}\right\}, & \text{for } q \neq q',\ r = r', \\ e^{-(\sigma^t_{pn}+\sigma^r_{pn})}, & \text{for } q \neq q',\ r \neq r', \end{cases} \tag{6.39}$$

where σ^t_{pn} and σ^r_{pn} are the phase noise variances at the transmitter and receiver oscillators, respectively. For a free-running oscillator, the phase noise process is modeled by a Brownian motion process with the difference between two realizations of the phase noise at time n and n' following a Normal-distributed random variable with zero mean and variance $4\pi f_{3\,\mathrm{dB}} T_s|n-n'|$ and $f_{3\,\mathrm{dB}}$ is a parameter describing the oscillator quality [17]. Then we have:

$$\mathbb{E}\left\{e^{j(\phi^s_{n',q}-\phi^s_{n,q})}\right\} = e^{-\frac{4\pi f_{3\,\mathrm{dB}} T_s|n-n'|}{2}}, \tag{6.40}$$

with T_s being the sample period. For PLL oscillator, the output phase noise is modeled as Ornstein-Uhlenbeck process [18].

For a shared-oscillator at the transmitter and the receiver,[2] the covariance of $e^{j\phi^s_{n,r,q}}$ and $e^{j\phi^s_{n',r',q'}}$ reduces to:

[2]One shared-oscillator is used between the antennas of one transceiver, but two different oscillators at the transmitter and the receiver are needed since they are located a two different transceivers.

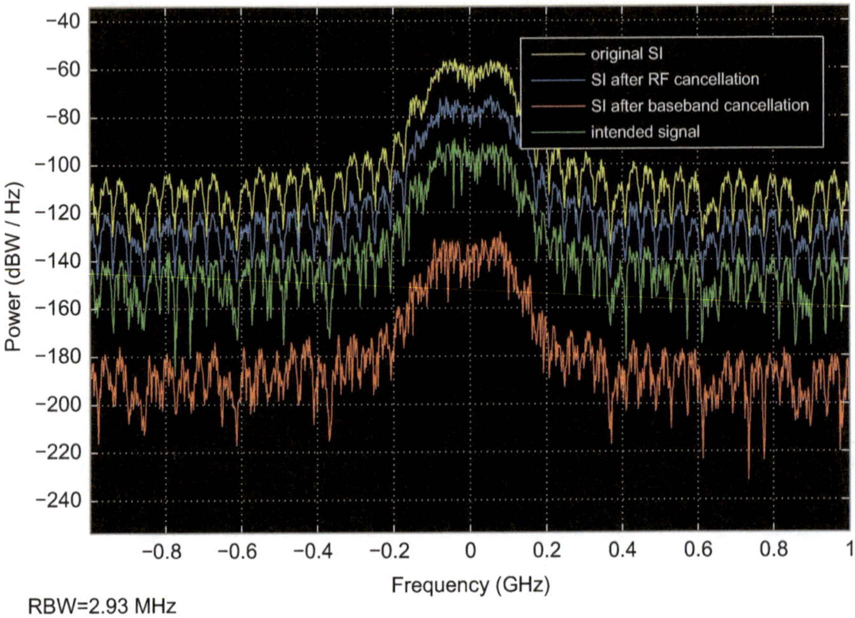

Fig. 6.13 SI spectrum with nonlinear cancellation

$$\mathbb{E}\left\{e^{j\phi^s_{n,r,q}}e^{-j\phi^s_{n',r',q'}}\right\} = \mathbb{E}\left\{e^{j(\phi_{n,r}-\phi_{n',r})}\right\}\mathbb{E}\left\{e^{j(\phi^s_{n',q}-\phi^s_{n,q})}\right\}, \tag{6.41}$$

which can be evaluated given the nature of the oscillator.

For independent transmit signal and phase noise process, it can be shown that:

$$\mathbb{E}\left\{\widetilde{\boldsymbol{S}}^d_t\overline{\boldsymbol{\Phi}}^s\overline{\boldsymbol{\Phi}}^{sH}\widetilde{\boldsymbol{S}}^{dH}_t\right\} = \mathbb{E}_s\left\{\widetilde{\boldsymbol{S}}^d_t\mathbb{E}_\phi\left\{\overline{\boldsymbol{\Phi}}^s\overline{\boldsymbol{\Phi}}^{sH}\right\}\widetilde{\boldsymbol{S}}^{dH}_t\right\}, \tag{6.42}$$

where $\mathbb{E}_s\{\cdot\}$ and $\mathbb{E}_\phi\{\cdot\}$ denote the expectation with respect to the unknown intended data and the phase noise, respectively. Considering the block structure of $\widetilde{\boldsymbol{S}}^d_t$, $\widetilde{\boldsymbol{S}}^d_t\mathbb{E}_\phi\left\{\overline{\boldsymbol{\Phi}}^s\overline{\boldsymbol{\Phi}}^{sH}\right\}\widetilde{\boldsymbol{S}}^{dH}_t$ is written as:

$$\begin{aligned}&\begin{pmatrix}\widetilde{\boldsymbol{S}}^d_1 & & \\ & \ddots & \\ & & \widetilde{\boldsymbol{S}}^d_{N_t}\end{pmatrix}\begin{pmatrix}\boldsymbol{R}_{1,1} & \dots & \boldsymbol{R}_{1,N_r} \\ \vdots & & \vdots \\ \boldsymbol{R}_{N_r,1} & \dots & \boldsymbol{R}_{N_r,N_r}\end{pmatrix}\begin{pmatrix}\widetilde{\boldsymbol{S}}^{dH}_1 & & \\ & \ddots & \\ & & \widetilde{\boldsymbol{S}}^{dH}_{N_t}\end{pmatrix}\\ &= \begin{pmatrix}\widetilde{\boldsymbol{S}}^d_1\boldsymbol{R}_{1,1}\widetilde{\boldsymbol{S}}^{dH}_1 & \dots & \widetilde{\boldsymbol{S}}^d_1\boldsymbol{R}_{1,N_r}\widetilde{\boldsymbol{S}}^{dH}_{N_r} \\ \vdots & & \vdots \\ \widetilde{\boldsymbol{S}}^d_{N_r}\boldsymbol{R}_{N_r,1}\widetilde{\boldsymbol{S}}^{dH}_1 & \dots & \widetilde{\boldsymbol{S}}^d_{N_r}\boldsymbol{R}_{N_r,N_r}\widetilde{\boldsymbol{S}}^{dH}_{N_r}\end{pmatrix},\end{aligned}$$

where $\boldsymbol{R}_{m,n} = \mathbb{E}\left\{\overline{\boldsymbol{\Phi}}^{s}_{m}\overline{\boldsymbol{\Phi}}^{sH}_{n}\right\}$ is a $NN_t \times NN_t$ matrix. Then, it can be shown that:

$$\widetilde{\boldsymbol{S}}^{d}_{m}\boldsymbol{R}_{m,n}\widetilde{\boldsymbol{S}}^{dH}_{n} = \sum_{q=1}^{N_t}\sum_{q'=1}^{N_t}\widetilde{\boldsymbol{S}}^{d}_{m,q}\boldsymbol{R}^{qq'}_{m,n}\widetilde{\boldsymbol{S}}^{dH}_{m,q}, \tag{6.43}$$

with $\boldsymbol{R}^{qq'}_{m,n} = \mathbb{E}\left\{\overline{\boldsymbol{\Phi}}^{s}_{m,q}\overline{\boldsymbol{\Phi}}^{sH}_{n,q'}\right\}$. By developing element (i,j) of the matrix $\widetilde{\boldsymbol{S}}^{d}_{m,q}\boldsymbol{R}^{qq'}_{m,n}\widetilde{\boldsymbol{S}}^{dH}_{m,q}$, we have:

$$\left[\widetilde{\boldsymbol{S}}^{d}_{m,q}\boldsymbol{R}^{qq'}_{m,n}\widetilde{\boldsymbol{S}}^{dH}_{m,q'}\right][i,j] = \sum_{l_1=0}^{L}\sum_{l_2=0}^{L} h^{s}_{m,q}(l_1)s^{d}_{q}(i-l_1)\boldsymbol{R}^{qq'}_{m,n}[i,j]h^{s*}_{n,q'}(l_2)s^{d*}_{q'}(j-l_2). \tag{6.44}$$

Since the intended signals are independently transmitted over the different antennas (i.e., $\mathbb{E}\left\{s^{d}_{q}(i)s^{d*}_{q'}(j)\right\} = 0$ if $q \neq q'$), it can be verified that $\mathbb{E}\left\{\left[\widetilde{\boldsymbol{S}}^{d}_{m,q}\boldsymbol{R}^{qq'}_{m,n}\widetilde{\boldsymbol{S}}^{dH}_{m,q}\right][i,j]\right\} = 0$ for $q \neq q'$. For $q = q'$, and noting that $\mathbb{E}\left\{s^{d}_{q}(i)s^{d*}_{q}(j)\right\} = 0$ if $i \neq j$, the element (i,j) in (6.44) can therefore be calculated as:

$$\mathbb{E}\left\{\left[\widetilde{\boldsymbol{S}}^{d}_{m,q}\boldsymbol{R}^{qq}_{m,n}\widetilde{\boldsymbol{S}}^{dH}_{m,q}\right][i,j]\right\} = \alpha^2 \sum_{l=\max(0,i-j)}^{\min(L,L+i-j)} h^{s}_{m,q}(l)\boldsymbol{R}^{qq}_{m,n}[i,j]h^{s*}_{n,q}(l). \tag{6.45}$$

Thus, combining (6.43) and (6.45), the expression of $\widetilde{\boldsymbol{R}}_{pn}$ is obtained with (6.39) for separate-oscillators and (6.41) for a shared-oscillator.

Appendix 3: Proof of Convergence

We prove in this appendix that the proposed iterative method converges to $(h^{rsi}_{r,q}e^{j\phi^{i}_{0,r,q}}, \Delta^{i}_{n,r,q})$ and $(h^{s}_{r,q}e^{j\phi^{s}_{0,r,q}}, \Delta^{s}_{n,r,q})$. Following the notations in Sect. 6.3, each phase noise process $\phi^{i}_{n,r,q}$ is represented as the sum of a constant term $\phi^{i}_{0,r,q}$ and a variable term $\Delta^{i}_{n,r,q}$. Small values of $\Delta^{i}_{n,r,q}$ and $\Delta^{s}_{n,r,q}$ satisfy $e^{j\Delta^{i}_{n,r,q}} \approx 1 + j\Delta^{i}_{n,r,q}$. Therefore, we have $\boldsymbol{\Phi} = \boldsymbol{\Phi}_0 + j\boldsymbol{\Delta}_t$ and the log-likelihood function (6.11) when considering the presence of phase noise can be rewritten as:

$$\mathscr{L}(\boldsymbol{h}) = -T\log|\boldsymbol{R}| - \sum_{t=1}^{T}\left(\boldsymbol{y}_t - \boldsymbol{D}_t\overline{\boldsymbol{h}}\right)^H \boldsymbol{R}^{-1}\left(\boldsymbol{y}_t - \boldsymbol{D}_t\overline{\boldsymbol{h}}\right) + \mathscr{F}(\boldsymbol{\Delta}_t, \boldsymbol{h}), \tag{6.46}$$

where the elements of $\overline{\boldsymbol{h}}$ are written as $h^{rsi}_{r,q}e^{j\phi^{i}_{0,r,q}}$ and $h^{s}_{r,q}e^{j\phi^{s}_{0,r,q}}$ and $\mathscr{F}(\boldsymbol{\Delta}_t, \boldsymbol{h})$ is a function of the channel coefficients and the variable part of the phase noise. The first iteration of the joint channel and phase noise estimation procedure is performed by:

$$\arg\max_{\bar{\boldsymbol{h}}} = -T\log|\boldsymbol{R}| - \sum_{t=1}^{T}\left(\boldsymbol{y}_t - \boldsymbol{D}_t\bar{\boldsymbol{h}}\right)^H \boldsymbol{R}^{-1}\left(\boldsymbol{y}_t - \boldsymbol{D}_t\bar{\boldsymbol{h}}\right), \tag{6.47}$$

and ignores the other terms containing the variable part of the phase noise. Thus, the problem in (6.47) returns an estimate of $\bar{\boldsymbol{h}}$, which will be considered as an estimate of the channel, leaving the variable terms $\boldsymbol{\Delta}_t$ to be estimated in the second step. As we iterate, the reference matrix $\boldsymbol{D}_t$ used to estimate the channel is rotated by the phase noise coefficients obtained in the previous iteration, which leads to the reduction of the contribution of $\mathscr{F}(\boldsymbol{\Delta}_t, \boldsymbol{h})$ in the log-likelihood function (6.46). Therefore, a better estimate of $\bar{\boldsymbol{h}}$ can be obtained, which results a better estimate of $\boldsymbol{\Delta}_t$ during the second step of the iteration. The iterative procedure guarantees a monotonic increase of the log-likelihood function through the set of re-estimation transformations. As a conclusion, the proposed algorithm converges to the point $(\bar{\boldsymbol{h}}, \boldsymbol{\Delta}_t)$ instead of $(\boldsymbol{h}, \boldsymbol{\Phi})$.

Appendix 4: Stochastic CRB

The CRB is defined as the inverse of the Fisher Information Matrix (FIM) [15, 19]. Following the derivations in [8], the real FIM can be formulated as:

$$\boldsymbol{J}_R(\boldsymbol{h}) = 2\begin{pmatrix} \Re(\boldsymbol{J}_{hh}) & -\Im(\boldsymbol{J}_{hh}) \\ \Im(\boldsymbol{J}_{hh}) & \Re(\boldsymbol{J}_{hh}) \end{pmatrix} + 2\begin{pmatrix} \Re(\boldsymbol{J}_{hh^*}) & -\Im(\boldsymbol{J}_{hh^*}) \\ \Im(\boldsymbol{J}_{hh^*}) & \Re(\boldsymbol{J}_{hh^*}) \end{pmatrix}, \tag{6.48}$$

where

$$\boldsymbol{J}_{hh}(i,j) = \left(\sum_{t=1}^{T} \boldsymbol{D}_t^H \boldsymbol{R}^{-1} \boldsymbol{D}_t\right)(i,j) + \operatorname{trace}\left(\boldsymbol{R}^{-1}\frac{\partial \boldsymbol{R}}{\partial \boldsymbol{h}^*(i)}\boldsymbol{R}^{-1}\frac{\partial \boldsymbol{R}}{\partial \boldsymbol{h}^*(j)}\right),$$

$$\boldsymbol{J}_{hh^*}(i,j) = \operatorname{trace}\left(\boldsymbol{R}^{-1}\frac{\partial \boldsymbol{R}}{\partial \boldsymbol{h}^*(i)}\boldsymbol{R}^{-1}\frac{\partial \boldsymbol{R}}{\partial \boldsymbol{h}^*(j)}\right). \tag{6.49}$$

The first derivative of $\boldsymbol{R}$ with respect to $\boldsymbol{h}^*(i)$ is:

$$\frac{\partial \boldsymbol{R}}{\partial \boldsymbol{h}^*(i)} = \begin{cases} \boldsymbol{0}, & \text{for } i = 1, \ldots, 4N_tN_r(L+1) \\ \alpha^2 \boldsymbol{H}^s \frac{\partial \boldsymbol{H}^s}{\partial \boldsymbol{h}^*(i)}, & \text{otherwise.} \end{cases} \tag{6.50}$$

The expression of the CRB depends on the specific realization of the channel. Therefore, we average the obtained CRB over a set of independent realizations of the channel coefficients. Note that in (6.50), we keep the dependence of the covariance matrix $\boldsymbol{R}$ on $\boldsymbol{h}^s$.

References

1. A. Masmoudi and T. Le-Ngoc, "A maximum-likelihood channel estimator for self-interference cancellation in full-duplex systems," *IEEE Trans. Vehicular Technology*, vol. 65, no. 7, pp. 5122–5132, July 2016.
2. ——, "Channel estimation and self-interference cancellation in full-duplex communication systems," *IEEE Trans. Vehicular Technology*, vol. 66, no. 1, pp. 321–334, Jan. 2017.
3. A. Sahai, G. Patel, C. Dick, and A. Sabharwal, "On the impact of phase noise on active cancelation in wireless full-duplex," *IEEE Trans. on Vehicular Technology*, vol. 62, no. 9, pp. 4494–4510, Nov 2013.
4. A. Masmoudi and T. Le-Ngoc, "Self-interference cancellation limits in full-duplex communication systems," in *Proc. IEEE Global Telecommun. Conf.*, Washington DC, USA, Dec. 2016.
5. V. Syrjala, M. Valkama, L. Anttila, T. Riihonen, and D. Korpi, "Analysis of oscillator phase-noise effects on self-interference cancellation in full-duplex OFDM radio transceivers," *IEEE Trans. on Wireless Comm.*, vol. 13, no. 6, pp. 2977–2990, June 2014.
6. E. Ahmed, A. Eltawil, and A. Sabharwal, "Self-interference cancellation with phase noise induced ICI suppression for full-duplex systems," in *Proc. IEEE Global Telecommun. Conf.*, Atlanta, Georgia, USA, Dec. 2013.
7. R. Li, A. Masmoudi, and T. Le-Ngoc, "Self-interference cancellation with phase-noise suppression in full-duplex systems," in *Proc. IEEE Pers. Indoor and Mobile Radio Commun.*, Hong Kong, China, Aug. 2015.
8. E. De Carvalho and D. T. Slock, "Cramer-Rao bounds for semi-blind, blind and training sequence based channel estimation," in *First IEEE Signal Processing Workshop on Signal Processing Advances in Wireless Communications*, Paris, France, Apr. 1997.
9. F. Wan, W.-P. Zhu, and M. Swamy, "A semiblind channel estimation approach for MIMO-OFDM systems," *IEEE Trans. Signal Process.*, vol. 56, no. 7, pp. 2821–2834, 2008.
10. H. Ochiai and H. Imai, "Performance analysis of deliberately clipped OFDM signals," *IEEE Trans. Comm.*, vol. 50, no. 1, pp. 89–101, 2002.
11. L. L. Scharf, *Statistical signal processing*. Addison-Wesley Reading, 1991, vol. 98.
12. A. Masmoudi and T. Le-Ngoc, "A maximum-likelihood channel estimator in MIMO full-duplex systems," in *Proc. IEEE Veh. Technol. Conf.*, Vancouver, British Columbia, Canada, Sep. 2014.
13. P. Stoica and A. Nehorai, "Performance study of conditional and unconditional direction-of-arrival estimation," *IEEE Trans. Acoust., Speech and Signal Process.*, vol. 38, no. 10, pp. 1783–1795, Oct 1990.
14. U. Mengali and A. N. D'Andrea, *Synchronization techniques for digital receivers*. Springer, 1997.
15. S. M. Kay, *Fundamentals of statistical signal processing, Volume 1: Estimation theory*. Prentice Hall, 1993.
16. S. Talwar, M. Viberg, and A. Paulraj, "Blind separation of synchronous co-channel digital signals using an antenna array. Part I: Algorithms," *IEEE Trans. Signal Process.*, vol. 44, no. 5, pp. 1184–1197, 1996.
17. D. Petrovic, W. Rave, and G. Fettweis, "Effects of phase noise on OFDM systems with and without PLL: Characterization and compensation," *IEEE Trans. on Comm.*, vol. 55, no. 8, pp. 1607–1616, Aug 2007.
18. A. Mehrotra, "Noise analysis of phase-locked loops," *IEEE Trans. Circuits and Systems I*, vol. 49, no. 9, pp. 1309–1316, Sep 2002.
19. A. Masmoudi, F. Bellili, S. Affes, and A. Stéphenne, "Closed-form expressions for the exact Cramer-Rao bounds of timing recovery estimators from BPSK, MSK and square-QAM transmissions," *IEEE Trans. on Sign. Process.*, vol. 59, no. 6, pp. 2474–2484, 2011.

Chapter 7
Active Signal Injection for Self-Interference-Cancellation

Abstract A common problem in full-duplex operation is the existence of a large self-interference (SI) at the receiver input, which prevents the proper detection of the intended signal. This chapter presents a new active signal injection (ASI) approach for full-duplex multiple-input multiple-output (MIMO) OFDM systems to reduce the SI prior to the receiver low-noise amplifier (LNA) and the analog-to-digital converter (ADC) to avoid overloading them while ensuring a proper signal detection. Basically, the ASI approach aims to add to the transmitted signal, an appropriate pre-cancelling signal that can reduce the SI at the receiver input of the same transceiver, but does not affect the detection process of the other intended receiver. To that end, two methods to realize the ASI approach are proposed and studied. In the first one, called tone reservation, the pre-cancelling signal only occupies some reserved subcarriers that are not used for data transmission, so that the subcarrier orthogonality can separate the transmitted data signal from the pre-cancelling signal in data detection. In the second method, called constellation relaxation, the data signal constellation points are dynamically extended to include the pre-cancelling signal in such a way that it minimizes the received SI but does not affect the intended data detection. The designs of the effective pre-cancelling signal in two methods are formulated as the optimization problems with derived solutions. A combination of both methods is also proposed to further enhance the SI-cancellation capability. Illustrative simulation results indicate that the proposed methods can offer a very good SI-cancellation performance with simple additional baseband digital signal processing implementation in the transmitter and without any required change in the receiver.

The cancellation methods proposed in Chaps. 4–6 are based on the time domain approach by creating a cancelling signal and subtracting it from the received signal. Other approaches use beamforming for SI-cancellation, usually referred as spatial cancellation. In these methods, the antenna patterns are adaptively shaped to mitigate the SI by creating a null space in the direction of the receiving antennas of the same transceiver.

While the previous methods can significantly reduce the SI, they suffer from various drawbacks which complicate their implementation. Practical implementation of the RF cancellation stage needs an additional RF chain, for every receive

T. Le-Ngoc, A. Masmoudi, *Full-Duplex Wireless Communications Systems*, Wireless Networks, DOI 10.1007/978-3-319-57690-9_7

antenna, which rapidly increases the complexity/cost of the MIMO system and raises compactness issues. Therefore, one may wonder if it is better to use the additional RF chains to transmit in a half-duplex fashion.

Moreover, spatial cancellation techniques sacrifice some of the available antennas for SI-cancellation [1, 2], and hence reduce the available multiplexing gain and transmission rate compared to using all available antennas for data transmission. Thus, the full antenna resources are under-utilized, which arises a natural question: Is full-duplex with under-utilized antenna resources better or worse than half-duplex using all available antennas for data transmission. If the channel state information between the two transceivers is available, an MMSE precoding technique can be deployed while maintaining the degree of freedom provided by MIMO systems [1]. In general, the precoders require complex optimization procedures and most of them are designed for flat frequency channel. Thus, they have to be implemented for each subcarrier of the OFDM signal in a separate manner. Null-space projection methods [1, 3] require that the SI channel is not full-rank to project the SI on the null space of the SI channel, a condition that is not always satisfied in practice. Accordingly, improving the SI-cancellation capability requires other solutions to be explored.

In this chapter, we resort to an entirely different approach for SI suppression/cancellation at the receiver input in full-duplex systems. The proposed approach, called active signal injection (ASI), intends to add an appropriately chosen *SI-cancelling* signal to the transmit signal to suppress/cancel the SI existing at the receiving antennas of the same transceiver for the following key potential advantages. The proposed ASI can be viewed as a spatial-domain SI-cancellation scheme, but without precoding to avoid the consequence of reducing the degree of freedom offered by MIMO systems. Furthermore, while sharing the same objective of the RF cancellation stage to suppress the SI at receiver input [10] (to avoid receiver overloading/saturation), the proposed ASI avoids the additional complexity and power consumption in SI-replica construction and subtraction required in a RF SI-cancellation stage.

When designing the SI-cancelling signal, we should keep in mind that the receiver is still able to decode the data-bearing signal without necessarily knowing the cancelling signal.

To that end, two ASI methods are presented. In the first one, called tone reservation, a small set of subcarriers are reserved to transmit the cancelling signal whose effects are seen in the time-domain. Since the subcarriers are orthogonal, the cancelling signal will not cause any distortion to the data-bearing signal. Fortunately, the problem of finding the values of these subcarriers can be formulated as a convex optimization problem for which we are able to obtain a closed-form solution. This method adds a small complexity at the transmitter, but does not require any additional operation at the receiver side.

The second ASI method is based on the observation that, for any received point belonging to the constellation boundary, the receiver is able to decode the transmitted symbol correctly [4]. Therefore, the constellation can be relaxed such that the transmit data plus the cancelling signal, at each subcarrier, still belongs to the boundary of the data point. This extra DoF is exploited to reduce the SI.

The constellation points are dynamically extended in an active manner to reduce the received SI. The method is labelled constellation relaxation since we are not restricted to a given number of points. These two methods are inspired from the solutions proposed in [5, 6] for the problem of peak-to-average power ratio (PAPR) in OFDM modulation systems and can also be combined to enhance the cancellation performance. For the PAPR problem, the objective is to reduce the peak of the OFDM symbol in the time domain by slightly changing the data on some or all subcarriers. While the objective of the proposed method is to minimize the combined signals from the transmit antennas at the receive antennas of the same transceiver. Thus we modify the techniques, originally designed for PAPR reduction, to the problem of reducing the received SI.

The remainder of this chapter is organized as follows. Section 7.1 presents the idea of the active signal injection cancellation structure. Section 7.2 presents the tone reservation method and Sect. 7.3 presents the constellation relaxation method. A combination of these two methods is presented in Sect. 7.4. Section 7.5 provides some simulation results and Sect. 7.6 presents the conclusion.

7.1 Novel Active Signal Injection for SI-Cancellation

In the following, we introduce the new ASI approach for SI-cancellation. The general formulation is as follows:

$$\begin{aligned}\bar{x}_q(n) &= x_q(n) + c_q(n) \\ &= \frac{1}{\sqrt{N}} \sum_{k=0}^{N-1} (X_q(k) + C_q(k)) e^{\frac{j2\pi kn}{N}},\end{aligned} \tag{7.1}$$

where the time-domain signal $c_q(n)$, or its equivalent frequency-domain representation $\boldsymbol{C}_q = \left[C_q(0), \ldots, C_q(N-1)\right]^T$, is injected at transmit antenna q and designed to reduce the SI. The bared signal $\bar{x}_q(n)$ denotes the signal that will be transmitted, i.e., the data-bearing signal $x_q(n)$ plus the cancelling signal $c_q(n)$. Figure 7.1 gives an illustrative representation of the proposed method. For this structure to be effective, the developed method should satisfy the following properties:

1. Choose $c_q(n)$ to reduce the SI at the receiver input of the same transceiver, after that the transmitted signal $\bar{x}_q(n)$ passes through the transmit chain and the SI channel.
2. The other intended receiver (of the other transceiver) should be able to efficiently decode the data-bearing signal $X_q(k)$ from the combined $X_q(k) + C_q(k)$ without necessarily knowing $C_q(k)$.
3. The injected signal $C_q(k)$ should not reduce the data rate significantly.

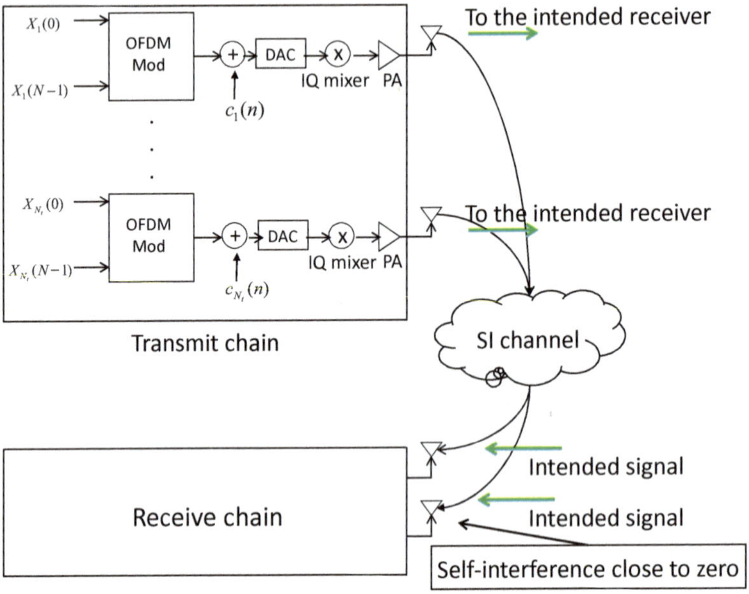

Fig. 7.1 Simplified diagram for the proposed ASI SI reduction method

The following methods are proposed to design the cancelling signal:

- Tone reservation: For which the data-bearing signal $X_q(k)$ and the injected signal $C_q(k)$ occupy disjoint subcarriers.
- Constellation relaxation: Where the symbol $X_q(k) + C_q(k)$ represented as a point in the complex plane belongs to the decision region of $X_q(k)$.

In the following sections, we give detail implementations of the proposed methods.

7.2 Tone Reservation for SI-Cancellation

The intended receiver must decode the values of $\boldsymbol{X}_q$ from the received vector $\boldsymbol{X}_q + \boldsymbol{C}_q$. In the tone reservation method, the two transceivers agree on reserving a small number of subcarriers for SI reduction. By constraining the injected signal $c_q(n)$ to lay in the reserved subcarriers, the data-bearing vector $\boldsymbol{X}_q$ can be easily separated from $\boldsymbol{C}_q$. Thus, the intended receiver does not need to know the value of $\boldsymbol{C}_q$. The signal $\boldsymbol{c}_q(n)$ is designed such that it has frequency support on the reserved subcarriers and its effect is seen in the time-domain. In the following, we show that the problem of finding the injected signal can be solved in a tractable manner.

Let $\mathscr{R}_q = \{i_1, \ldots, i_{R_q}\}$ represent the set of subcarrier's index that are reserved for SI-cancellation at the transmitting antenna q, where R_q is the number of reserved subcarriers. For simplicity, we assume that $R_q = R$ for all antennas $q = 1, \ldots, N_t$. Calling $\mathscr{R}_q^c$ the complement of $\mathscr{R}_q$ in the set $\{0, \ldots, N-1\}$, we have:

$$X_q(k) + C_q(k) = \begin{cases} X_q(k), & \text{if } k \in \mathscr{R}_q^c, \\ C_q(k), & \text{if } k \in \mathscr{R}_q. \end{cases} \tag{7.2}$$

With this choice of $\boldsymbol{C}_q$, the demodulation is not affected by the cancelling signal. The only change at the receiver is to only decode the signal in the subcarriers $k \in \mathscr{R}_q^c$. It follows that the OFDM signal at transmit antenna q can be written as:

$$\overline{x}_q(n) = \frac{1}{\sqrt{N}} \sum_{k \in \mathscr{R}_q^c} X_q(k) e^{\frac{j2\pi kn}{N}} + \frac{1}{\sqrt{N}} \sum_{k \in \mathscr{R}_q} C_q(k) e^{\frac{j2\pi kn}{N}}. \tag{7.3}$$

As discussed in the previous section, the transmit signal $\overline{x}_q(n)$ is passed through an IQ mixer and becomes [7, 8]:

$$\overline{x}_q^{IQ}(n) = \overline{x}_q(n) + g_{2,q}\overline{x}_q^*(n). \tag{7.4}$$

To simplify the development, we assume negligible PA induced impairments while considering the PA impairments and ADC quantization noise in the simulations. We mention that the presence of the SI-induced from the PA impairment does not increase the quantization noise and that the PA impairments will be suppressed by the baseband SI-cancellation stage as detailed in Chaps. 4–6. The received signal at the Rx stream r is:

$$y_r(n) = \sum_{q=1}^{N_t} \sum_{l=0}^{L} \left(h_{r,q}^i(l)\overline{x}_q^{IQ}(n-l) + h_{r,q}^s(l)s_q(n-l) \right) + w_r(n), \tag{7.5}$$

where $h_{r,q}^i(l)$ is the SI channel between the Tx stream q and Rx stream r of the same transceiver and $h_{r,q}^s(l)$ is the intended channel of the link from Tx stream q of the other intended transmitter to Rx stream r. Our goal is to find the set of vectors $\boldsymbol{C}_q$, for $q = 1, \ldots, N_t$, that minimize the received SI, given by the term:

$$y_r^{\text{SI}}(n) = \sum_{q=1}^{N_t} \sum_{l=0}^{L} h_{r,q}^i(l)\overline{x}_q^{IQ}(n-l), \tag{7.6}$$

under the constraint that $C_q(k) = 0$ for $k \notin \mathscr{R}_q$. That is, the problem at hand is formulated as:

$$\begin{aligned} &\min_{\boldsymbol{C}_q,\ q=1,\ldots,N_t} \sum_{r=1}^{N_r} \|\boldsymbol{y}_r^{\text{SI}}\|_2^2, \\ &\text{such that } C_q(k) = 0, \ \text{for } k \notin \mathscr{R}_q,\ q = 1, \ldots, N_t, \end{aligned} \tag{7.7}$$

where $\boldsymbol{y}_r^{\mathrm{SI}} = \left[y_r^{\mathrm{SI}}(0), \ldots, y_r^{\mathrm{SI}}(N-1)\right]^T$ and the condition on the reserved subcarriers is expressed in the constraint. Solving the problem[1] as it is presented in (7.7) seems to be complicated. Therefore, we first include the constraint in the objective function to obtain an unconstrained minimization problem, easier to solve. First, note that the fast Fourier transform (FFT) coefficients of $\boldsymbol{x}_q^*$, denoted $\widetilde{\boldsymbol{X}}_q$, are related to the FFT coefficients of $\boldsymbol{x}_q$ as:

$$\begin{aligned}
\widetilde{X}_q(k) &= \sum_{n=0}^{N-1} x^*(n) e^{-\frac{j2\pi nk}{N}} \\
&= \left(\sum_{n=0}^{N-1} x(n) e^{\frac{j2\pi nk}{N}}\right)^* \\
&= \left(\sum_{n=0}^{N-1} x(n) e^{-\frac{j2\pi n(N-k)}{N}}\right)^* \\
&= \begin{cases} X_q^*(k), & \text{for } k = 0 \text{ or } \frac{N}{2}, \\ X_q^*(N-k), & \text{for } k \neq 0 \text{ and } \frac{N}{2}. \end{cases}
\end{aligned} \tag{7.8}$$

Before further development, let $\mathcal{Q}_k = \{q_1, \ldots, q_n\}$ denote the subset of antennas that have reserved subcarrier k for SI-cancellation[2] (i.e., $q_n \in \mathcal{Q}_k$ if $k \in \mathcal{R}_{q_n}$). For a fixed subcarrier k, we distinguish three cases:

1. If the antenna index $q \in \mathcal{Q}_k$, then $X_q(k) = 0$ and $C_q(k) \neq 0$.
2. If the antenna index $q \in \mathcal{Q}_{N-k}$, then $X_q(N-k) = 0$ and $C_q(N-k) \neq 0$.
3. If the antenna index $q \notin \mathcal{Q}_k \cup \mathcal{Q}_{N-k}$, then $X_q(k) \neq 0$, $X_q(N-k) \neq 0$, $C_q(k) = 0$ and $C_q(N-k) = 0$.

Therefore, the received SI signal $Y_r^{\mathrm{SI}}(k)$ at antenna r in the frequency domain can be written as:

$$\begin{aligned}
Y_r^{\mathrm{SI}}(k) &= \sum_{q=1}^{N_t} d_{rq}(k)\big(X_q(k) + g_{2,q}X_q^*(N-k) + C_q(k) + g_{2,q}C_q^*(N-k)\big) \\
&= \sum_{q\in\mathcal{Q}_k} d_{rq}(k)C_q(k) + \sum_{q\in\mathcal{Q}_{N-k}} d_{rq}(k)g_{2,q}C_q^*(N-k) + Z_{r,k},
\end{aligned} \tag{7.9}$$

[1]The original problem is of the form $\min_{C_q,\, q=1,\ldots,N_t} \max_r ||\boldsymbol{y}_r^{\mathrm{SI}}||^2$, such that $C_q(k) = 0$, for $k \notin \mathcal{R}_q$, $q = 1, \ldots, N_t$, for which a closed-form solution, may not be possible to derive. Therefore, we resort to the formulation in (7.7) leading to a closed-form solution and the SI reduction over all receive antennas.

[2]The number of elements in $\mathcal{Q}_k$ may change from one subcarrier to another.

where $d_{rq}(k)$ is the SI channel response at subcarrier k and $Z_{r,k}$ is the data-dependent term given by:

$$Z_{r,k} = \sum_{q\in\mathscr{Q}_k} d_{rq}(k)g_{2,q}X_q^*(N-k) + \sum_{q\in\mathscr{Q}_{N-k}} d_{rq}(k)X_q(k) + \sum_{q\notin\mathscr{Q}_k\cup\mathscr{Q}_{N-k}} d_{rq}(k)(X_q(k)+g_{2,q}X_q^*(N-k)). \quad (7.10)$$

Thus, using the Parseval equality, $||\mathbf{y}_r^{\mathrm{SI}}||_2^2$ is equivalent to its following frequency-domain representation:

$$\begin{aligned} ||\mathbf{y}_r^{\mathrm{SI}}||_2^2 &= \sum_{k=0}^{N-1} |y_r^{\mathrm{SI}}(k)|^2 \\ &= \sum_{k=0}^{N-1} |Y_r^{\mathrm{SI}}(k)|^2 \\ &= \sum_{k=0}^{N-1} \left| \sum_{q\in\mathscr{Q}_k} d_{rq}(k)C_q(k) + \sum_{q\in\mathscr{Q}_{N-k}} d_{rq}(k)g_{2,q}C_q^*(N-k) + Z_{r,k} \right|^2. \quad (7.11) \end{aligned}$$

We now manipulate $||\mathbf{y}_r^{\mathrm{SI}}||_2^2$ in (7.11) in order to make the problem of reducing the SI in (7.7) easy to solve. In the following, we distinguish two cases.

First, for $k=0$ and $k=N/2$, we have $\widetilde{X}_q(k) = X_q^*(k)$. Noting that $\mathscr{Q}_k = \mathscr{Q}_{N-k}$ for $k=0$ and $k=N/2$, $|Y_r^{\mathrm{SI}}(k)|^2$ can be simply written as:

$$\begin{aligned} |Y_r^{\mathrm{SI}}(k)|^2 &= \left| \sum_{q\in\mathscr{Q}_k} \left(d_{rq}(k)C_q(k) + d_{rq}(k)g_{2,q}C_q^*(k)\right) + Z_{r,k} \right|^2 \\ &= \left| \mathbf{C}_{\mathscr{Q}_k}^T \mathbf{d}_{r,\mathscr{Q}_k} + \mathbf{C}_{\mathscr{Q}_k}^H \widetilde{\mathbf{d}}_{r,\mathscr{Q}_k} + Z_{r,k} \right|^2, \quad (7.12) \end{aligned}$$

where $\mathbf{C}_{\mathscr{Q}_k}$ is the vector containing the symbols from antennas that have a reserved subcarrier at position k (i.e., $\mathbf{C}_{\mathscr{Q}_k}(q) = C_q(k)$ for $q \in \mathscr{Q}_k$), $\mathbf{d}_{r,\mathscr{Q}_k} = \left[d_{rq_1}(k), \ldots, d_{rq_Q}(k)\right]^T$ and $\widetilde{\mathbf{d}}_{r,\mathscr{Q}_k} = \left[g_{2,q_1}d_{rq_1}(k), \ldots, g_{2,q_Q}d_{rq_Q}(k)\right]^T$, for $\{q_1, \ldots, q_Q\} \subset \mathscr{Q}_k$. By developing $|Y_r^{\mathrm{SI}}(k)|^2$, we obtain the following matricial expression:

$$\begin{aligned} |Y_r^{\mathrm{SI}}(k)|^2 = \; & \mathbf{C}_{\mathscr{Q}_k}^T \mathbf{M}_{d,d^*}^k \mathbf{C}_{\mathscr{Q}_k}^* + \mathbf{C}_{\mathscr{Q}_k}^T \mathbf{M}_{d,\widetilde{d}^*}^k \mathbf{C}_{\mathscr{Q}_k} + \mathbf{C}_{\mathscr{Q}_k}^H \mathbf{M}_{\widetilde{d},d^*}^k \mathbf{C}_{\mathscr{Q}_k}^* + \mathbf{C}_{\mathscr{Q}_k}^H \mathbf{M}_{\widetilde{d},\widetilde{d}^*}^k \mathbf{C}_{\mathscr{Q}_k} \\ & + \mathbf{d}_{r,\mathscr{Q}_k}^T \mathbf{C}_{\mathscr{Q}_k} Z_{r,k}^* + \widetilde{\mathbf{d}}_{r,\mathscr{Q}_k}^T \mathbf{C}_{\mathscr{Q}_k}^* Z_{r,k}^* + Z_{r,k}\left(\mathbf{d}_{r,\mathscr{Q}_k}^H \mathbf{C}_{\mathscr{Q}_k}^* + \widetilde{\mathbf{d}}_{r,\mathscr{Q}_k}^H \mathbf{C}_{\mathscr{Q}_k} + Z_{r,k}^*\right), \quad (7.13) \end{aligned}$$

where the matrix $\boldsymbol{M}^k_{d,d*}$ is defined as:

$$\boldsymbol{M}^k_{d,d*} = \boldsymbol{d}_{r,\mathcal{Q}_k}\boldsymbol{d}^H_{r,\mathcal{Q}_k}. \tag{7.14}$$

In the same way, we also define $\boldsymbol{M}^k_{d,\widetilde{d}*} = \boldsymbol{d}_{r,\mathcal{Q}_k}\widetilde{\boldsymbol{d}}^H_{r,\mathcal{Q}_k}$, $\boldsymbol{M}^k_{\widetilde{d},d*} = \widetilde{\boldsymbol{d}}_{r,\mathcal{Q}_k}\boldsymbol{d}^H_{r,\mathcal{Q}_k}$ and $\boldsymbol{M}^k_{\widetilde{d},\widetilde{d}*} = \widetilde{\boldsymbol{d}}_{r,\mathcal{Q}_k}\widetilde{\boldsymbol{d}}^H_{r,\mathcal{Q}_k}$. In (7.13), the vector $\boldsymbol{C}_{\mathcal{Q}_k}$ appears in many positions by taking its transpose or its conjugate transpose every time. To factorize all these terms, we develop the expression in (7.13) in term of the real and imaginary parts of its different elements. By introducing the extended vector $\widetilde{\boldsymbol{C}}_{\mathcal{Q}_k} = \left[\Re\{\boldsymbol{C}^T_{\mathcal{Q}_k}\}\ \Im\{\boldsymbol{C}^T_{\mathcal{Q}_k}\}\right]^T$, $\left|Y^{\mathrm{SI}}_r(k)\right|^2$ is rewritten as:

$$\left|Y^{\mathrm{SI}}_r(k)\right|^2 =$$

$$\widetilde{\boldsymbol{C}}^T_{\mathcal{Q}_k}\underbrace{\left(\begin{pmatrix}\boldsymbol{M}^k_{d,d*} & -j\boldsymbol{M}^k_{d,d*}\\ j\boldsymbol{M}^k_{d,d*} & \boldsymbol{M}^k_{d,d*}\end{pmatrix}+\begin{pmatrix}\boldsymbol{M}^k_{d,\widetilde{d}*} & j\boldsymbol{M}^k_{d,\widetilde{d}*}\\ j\boldsymbol{M}^k_{d,\widetilde{d}*} & -\boldsymbol{M}^k_{d,\widetilde{d}*}\end{pmatrix}+\begin{pmatrix}\boldsymbol{M}^k_{\widetilde{d},d*} & -j\boldsymbol{M}^k_{\widetilde{d},d*}\\ -j\boldsymbol{M}^k_{\widetilde{d},d*} & -\boldsymbol{M}^k_{\widetilde{d},d*}\end{pmatrix}+\begin{pmatrix}\boldsymbol{M}^k_{\widetilde{d},\widetilde{d}*} & j\boldsymbol{M}^k_{\widetilde{d},\widetilde{d}*}\\ -j\boldsymbol{M}^k_{\widetilde{d},\widetilde{d}*} & \boldsymbol{M}^k_{\widetilde{d},\widetilde{d}*}\end{pmatrix}\right)}_{\boldsymbol{M}_{r,k}}\widetilde{\boldsymbol{C}}_{\mathcal{Q}_k}$$

$$+\underbrace{2\left[\Re\{\boldsymbol{d}^T_{r,\mathcal{Q}_k}Z^*_{r,k} + \widetilde{\boldsymbol{d}}^T_{r,\mathcal{Q}_k}Z^*_{r,k}\}\ \Im\{\boldsymbol{d}^H_{r,\mathcal{Q}_k}Z_{r,k} + \widetilde{\boldsymbol{d}}^T_{r,\mathcal{Q}_k}Z^*_{r,k}\}\right]}_{\boldsymbol{v}^T_{r,k}}\widetilde{\boldsymbol{C}}_{\mathcal{Q}_k} + |Z_{r,k}|^2. \tag{7.15}$$

Finally, we obtain the following compact form:

$$\left|Y^{\mathrm{SI}}_r(k)\right|^2 = \widetilde{\boldsymbol{C}}^T_{\mathcal{Q}_k}\boldsymbol{M}_{r,k}\widetilde{\boldsymbol{C}}_{\mathcal{Q}_k} + \boldsymbol{v}^T_{r,k}\widetilde{\boldsymbol{C}}_{\mathcal{Q}_k} + |Z_{r,k}|^2. \tag{7.16}$$

For $k \neq 0$ and $k \neq N/2$ and using the fact that $\widetilde{C}_q(k) = C^*_q(N-k)$, $\left|Y^{\mathrm{SI}}_r(k)\right|^2$ is written as:

$$\left|Y^{\mathrm{SI}}_r(k)\right|^2 = \left|\sum_{q\in\mathcal{Q}_k} d_{rq}(k)C_q(k) + \sum_{q\in\mathcal{Q}_{N-k}} d_{rq}(k)g_{2,q}C^*_q(N-k) + Z_{r,k}\right|^2. \tag{7.17}$$

As shown in Appendix 1, $\left|Y^{\mathrm{SI}}_r(k)\right|^2$ can be written in a similar matricial form as for $k = 0$ and $k = N/2$ in (7.16). Thus, the constrained optimization problem in (7.7) is equivalent to the following unconstrained problem:

$$\min_{\widetilde{\boldsymbol{C}}} \widetilde{\boldsymbol{C}}^T\boldsymbol{M}\widetilde{\boldsymbol{C}} + \boldsymbol{v}^T\widetilde{\boldsymbol{C}} + \sum_{r=1}^{N_r}\sum_{k=0}^{N-1}|Z_{r,k}|^2, \tag{7.18}$$

where $\boldsymbol{M}$ and $\boldsymbol{v}$ are defined in (7.43) and the constraints on the reserved subcarriers in (7.7) are now included in the objective function in (7.18). The problem of minimizing the quadratic function in (7.18) is solved in a closed-form as:

$$\widetilde{C} = -M^{-1} v, \tag{7.19}$$

where we show in Appendix 2 that the matrix M is invertible.

The choice of the reserved subcarriers is an important issue for this method. The natural choice is to select $\mathscr{R}_q$ such that the SI is minimized. This is an NP-hard problem since we need to check all possible discrete sets $\mathscr{R}_q$ and cannot be solved for large values of subcarriers N. On the other hand, subcarriers with relatively low SNR, which are unused or carry low-rate data, can be instead reserved for SI reduction. This requires the reserved set $\mathscr{R}_q$ to be sent to the receiver. In this work, the reserved subcarriers are selected periodically over the frequency band while efficient algorithms to improve the selected subcarriers are left for future work.

One other important issue is the number of reserved subcarriers. Increasing the number of reserved subcarriers would improve the cancellation performance as we have more freedom to design the injected signal, but it also decreases the amount of transmitted data. Moreover, since the subcarriers are orthogonal, we need to reserve the whole interval $[0, N-1]$ to reduce the SI. Therefore, we distribute the reserved subcarriers over the transmit antennas. Considering that the signal image creates interaction between two subcarriers, a minimum of $\frac{N}{2N_t}$ reserved subcarriers are needed to cover the whole subcarriers when the reserved subcarriers subsets $\mathscr{R}_q$ at the different transmit antennas are disjoint.

To implement the proposed method, the transceiver is expected to know the SI channel. Therefore, a half-duplex transmission period is needed at the beginning to estimate the SI channel. This is a common assumption used to implement the RF cancellation [9–11] and the spatial cancellation [1, 2].

7.3 Constellation Relaxation for SI-Cancellation

In the constellation relaxation method, each symbol is allowed to be mapped to a point in a specific set of the complex plane. Our purpose is to reduce the SI by judiciously choosing the appropriate constellation points from the allowable set. That is, considering a 4-QAM modulation in each subcarrier, there is 4 points laying in each quadrant of the complex plane. The decision regions are the four quadrants bounded by the real and imaginary axes and an error occurs when the additive noise[3] translates the received sample into another quadrant. Any point that is farther from the decision boundaries than the actual constellation point will not affect the decision performance. Thus, it is possible to transmit any point within the quarter-plane outside the actual constellation point. This idea is illustrated in Fig. 7.2 where the gray region represents the possible extension of the corresponding constellation point. The injected signal $C_q(k)$ in the frequency-domain will be

[3]Assuming white Gaussian noise.

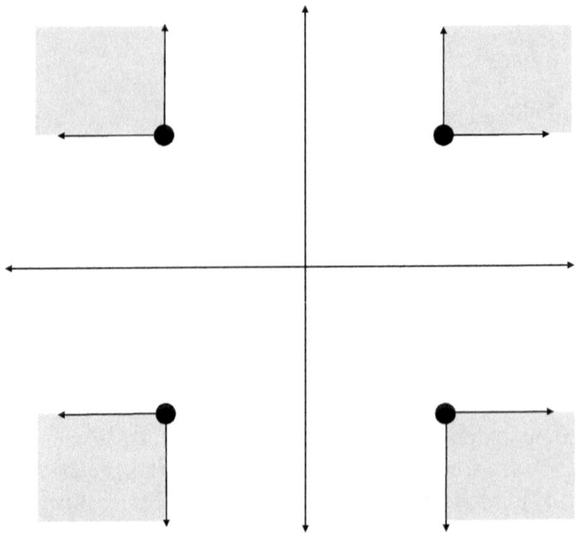

Fig. 7.2 Illustration of the constellation relaxation principle for 4-QAM. The *gray regions* represent the extension region for the corresponding point

chosen to only move the actual point in its corresponding gray region. That is, the point $X_q(k) + C_q(k)$ is still located in the decision region of $X_q(k)$. If properly adjusted, the combination of this injected signal can reduce the received SI.

This idea can be generalized to other higher-order M-QAM. The exterior points can be extended in the same way as for the 4-QAM while the interior points do not have the flexibility to be changed since they are limited from all sides. Figure 7.3 shows the extension process for 16-QAM constellation. For M-PSK constellations, the points lie on the outer boundaries and are extended in a region parallel to the decision boundary to maintain the minimum distance between the points. All these extra degrees of freedom are next exploited to reduce the SI.

For the constellation relaxation method, the problem of reducing the SI can be formulated as:

$$\min_{\boldsymbol{C}_q,\ q=1,\ldots,\ N_t} \sum_{r=1}^{N_r} ||\boldsymbol{y}_r^{\mathrm{SI}}||_2^2, \tag{7.20}$$

such that the additive signal $\boldsymbol{C}_q$, for $q = 1, \ldots, N_t$, satisfies the extension constraint available for every constellation point. We now look for a practical method to solve the problem in (7.20). We start by manipulating the cost function to highlight the variable to optimize. Unlike the tone reservation method, the injected signal has support over the whole frequency slot. Thus the OFDM signal to be transmitted at antenna q, before it goes through the IQ mixer, is given by:

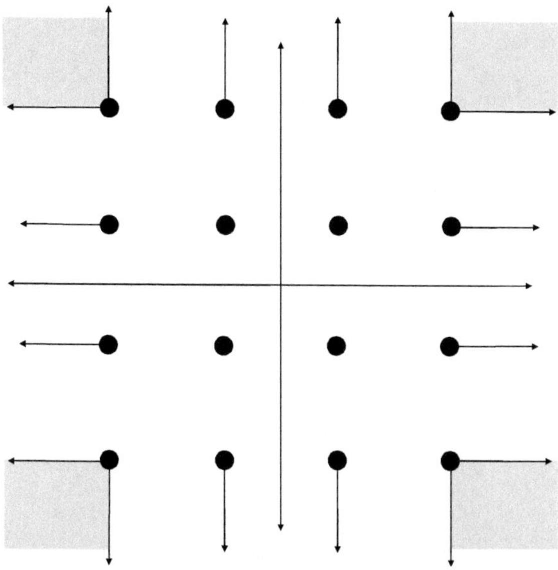

Fig. 7.3 Illustration of the constellation relaxation principle for 16-QAM. The *gray regions* represent the extension region for the corner points and the *arrows* represent the possible extension of the side points over one dimension

$$\overline{\boldsymbol{x}}_q(n) = \frac{1}{\sqrt{N}} \sum_{k=0}^{N-1} (X_q(k) + C_q(k)) e^{\frac{j2\pi kn}{N}}. \tag{7.21}$$

Then the received SI signal at the input of the receive antenna r is given by:

$$y_r^{\mathrm{SI}}(n) = \sum_{q=1}^{N_t} \sum_{l=0}^{L} h_{r,q}^{i}(l) \overline{\boldsymbol{x}}_q^{IQ}(n-l), \tag{7.22}$$

Applying again the Parseval's theorem $||\boldsymbol{y}_r^{\mathrm{SI}}||_2^2 = ||\boldsymbol{Y}_r^{\mathrm{SI}}||_2^2$, we have:

$$\begin{aligned} ||\boldsymbol{y}_r^{\mathrm{SI}}||_2^2 &= \sum_{k=0}^{N-1} |Y_r^{\mathrm{SI}}(k)|^2 \\ &= \sum_{q=1}^{N_t} \left| d_{rq}(k)(C_q(k) + g_{2,q}\widetilde{C}_q(k)) + W_{r,k} \right|^2, \end{aligned} \tag{7.23}$$

where

$$W_{r,k} = \sum_{q=1}^{N_t} d_{rq}(k) \left(X_q(k) + g_{2,q}\widetilde{X}_q(k) \right). \tag{7.24}$$

Note that the summation in (7.11) is performed over a set of selected antennas $\mathcal{Q}_k$ and $\mathcal{Q}_k^c$ for every subcarrier, while in (7.23), the summation is performed over all the transmitting antennas. Therefore, if we consider that $\mathcal{Q}_k = \mathcal{Q}_k^c = \{1, \ldots, N_t\}$, $\left|Y_r^{\text{SI}}(k)\right|^2$ is simply obtained following the same approach as in Sect. 7.2:

$$\left|Y_r^{\text{SI}}(k)\right|^2 = \widetilde{\boldsymbol{C}}_k^T \boldsymbol{N}_{r,k} \widetilde{\boldsymbol{C}}_k + \boldsymbol{u}_{r,k}^T \widetilde{\boldsymbol{C}}_k + |W_{r,k}|^2, \tag{7.25}$$

where $\boldsymbol{N}_{r,k}$ and $\boldsymbol{u}_{r,k}$ are built in the same as $\boldsymbol{M}_{r,k}$ and $\boldsymbol{v}_{r,k}$ in (7.16) for $k = 0$ or $k = N/2$, or as $\boldsymbol{M}_{r,k}$ and $\boldsymbol{v}_{r,k}$ in (7.37) given in Appendix 1. Thus we have:

$$||\boldsymbol{Y}_r^{\text{SI}}||_2^2 = \widetilde{\boldsymbol{C}}^T \boldsymbol{N}_r \widetilde{\boldsymbol{C}} + \boldsymbol{u}_r^T \widetilde{\boldsymbol{C}} + \sum_{k=0}^{N-1} |W_{r,k}|, \tag{7.26}$$

where $\boldsymbol{N}_r$ is the block diagonal matrix given by:

$$\boldsymbol{N}_r = \begin{pmatrix} \boldsymbol{N}_{r,1} & \boldsymbol{0} & \ldots & \boldsymbol{0} \\ \boldsymbol{0} & \boldsymbol{N}_{r,2} & & \vdots \\ \vdots & & \ddots & \boldsymbol{0} \\ \boldsymbol{0} & \ldots & \boldsymbol{0} & \boldsymbol{N}_{r,\frac{N}{2}} \end{pmatrix}, \tag{7.27}$$

$\boldsymbol{u}_r = \left[\boldsymbol{u}_{r,1}^T, \ldots, \boldsymbol{u}_{r,\frac{N}{2}}^T\right]^T$ depends on the SI channel and the transmitted data and $\widetilde{\boldsymbol{C}} = \left[\widetilde{\boldsymbol{C}}_0^T, \ldots, \widetilde{\boldsymbol{C}}_{\frac{N}{2}}^T\right]^T$. Then, the cost function in (7.20) is equivalent to the following quadratic function:

$$\sum_{r=1}^{N_r} ||\boldsymbol{Y}_r^{\text{SI}}||_2^2 = \widetilde{\boldsymbol{C}}^T \underbrace{\sum_{r=1}^{N_r} \boldsymbol{N}_r}_{\boldsymbol{N}} \widetilde{\boldsymbol{C}} + \underbrace{\sum_{r=1}^{N_r} \boldsymbol{u}_r^T}_{\boldsymbol{u}^T} \widetilde{\boldsymbol{C}} + \sum_{k=0}^{N-1} \sum_{r=1}^{N_r} W_{r,k}. \tag{7.28}$$

To have a tractable formulation of the constraint, consider for instance 4-QAM modulated signal. Then the problem can be mathematically formulated as [4]:

$$\begin{aligned} &\min_{\widetilde{\boldsymbol{C}}} \widetilde{\boldsymbol{C}}^T \boldsymbol{N} \widetilde{\boldsymbol{C}} + \boldsymbol{u}^T \widetilde{\boldsymbol{C}}, \\ &\text{such that} \quad -\operatorname{diag}\left\{\widetilde{\boldsymbol{X}}\right\} \widetilde{\boldsymbol{C}} \preceq \boldsymbol{0}, \end{aligned} \tag{7.29}$$

where $\preceq$ compares two vectors element by element and diag $\{\widetilde{\boldsymbol{X}}\}$ returns a diagonal matrix whose diagonal elements are the elements of the vector $\widetilde{\boldsymbol{X}}$ obtained by rearranging the transmitted symbols in the same way as $\widetilde{\boldsymbol{C}}$. The constraint in (7.29) restricts each element of $\widetilde{\boldsymbol{C}}$ to have the same sign as the corresponding element of $\widetilde{\boldsymbol{X}}$ (i.e., $\widetilde{\boldsymbol{X}}(n)$ and $\widetilde{\boldsymbol{C}}(n)$ have the same sign). As a consequence, every element is allowed to extend only in the allocated region. The attractive feature of the resulting problem in (7.29) is that the constraints are now linear and the cost function is quadratic, which makes the problem easier to solve using quadratic programming [12].

The constraints in (7.29) have to be adapted depending on the modulation scheme. Here, we present a procedure to change the constraints for 16-QAM and the same idea can be applied to most of the modulations. As illustrated in Fig. 7.3, the 4 interior points cannot be extended while the side points can be extended only in one dimension (either in the real or imaginary part). On the other hand, the 4 points in the corner can be extended in both dimensions. It follows that the injected signal $C_q(k)$ at subcarriers whose symbols $X_q(k)$ are at the interior points should be equal to zero. This requires adding an additional constraint to (7.29). Let $\boldsymbol{E}$ be a matrix whose elements are zeros except for the diagonal positions where $X_q(k)$ is an interior point. Thus the constraint $\boldsymbol{E}\widetilde{\boldsymbol{C}} = \boldsymbol{0}$ forces the injected signal $\widetilde{\boldsymbol{C}}$ to satisfy the constraint for the 16-QAM. Same principle is applied for the side points by making either the real or imaginary part of $C_q(k)$ equal to zero. With these additional constraints, the problem can still be solved using quadratic programming.

We should mention that, when $\boldsymbol{C}_q(k) \neq \boldsymbol{0}$, $\boldsymbol{X}_q(k) + \boldsymbol{C}_q(k)$ has more energy than the original signal $\boldsymbol{X}_q(k)$. This power increase can result in reducing the SNR margin. To control the power increase, the amplitude of $C_q(k)$ is constrained to be lower than a given value. We show through simulations that a good SI reduction is obtained with the expense of 1 dB power increase at the transmitter. Finally, the problem to solve, for 16-QAM constellation, is given by:

$$\begin{aligned}
&\min_{\widetilde{\boldsymbol{C}}} \widetilde{\boldsymbol{C}}^T \boldsymbol{N} \widetilde{\boldsymbol{C}} + \boldsymbol{u}^T \widetilde{\boldsymbol{C}}, \\
&\text{such that} \quad - \operatorname{diag}\{\widetilde{\boldsymbol{X}}\} \widetilde{\boldsymbol{C}} \preceq \boldsymbol{0}, \\
&\qquad\qquad\quad \boldsymbol{E}\widetilde{\boldsymbol{C}} = \boldsymbol{0}, \\
&\qquad\qquad\quad \widetilde{\boldsymbol{C}} \preceq L_{up}\boldsymbol{1}, \\
&\qquad\qquad\quad \widetilde{\boldsymbol{C}} \succeq L_{low}\boldsymbol{1},
\end{aligned} \tag{7.30}$$

where L_{up} and L_{low} represent the upper and lower bounds that limit the values of $\widetilde{\boldsymbol{C}}$ and $\boldsymbol{1}$ is the all-one vector. To keep the symmetry of the transmitted signal, we usually choose $L_{up} = -L_{low}$. In this case, the maximum power increase $P_{increase}$ is:

$$P_{increase} = 2L_{up}^2, \tag{7.31}$$

and the procedure is adapted to 4-QAM modulation by ignoring the constraint $\boldsymbol{E}\widetilde{\boldsymbol{C}} = \boldsymbol{0}$ in (7.30).

7.4 Combination of Both Methods

The tone reservation method needs to sacrifice some subcarriers to carry on the cancelling signal, while the constellation relaxation method increases the average transmit power. To further enhance the SI reduction while saving resources, we combine the two previous methods. In this case, we distinguish between two sets of subcarriers. One set contains the reserved subcarriers which are not constrained by the transmitted data, and one set of subcarriers contains both the data-bearing signal and the injected signal where the injected signal is chosen according to the rule described in the previous subsection. To that end, the constraints in (7.30) are modified to take into account the reserved subcarriers. Let $\widetilde{\boldsymbol{C}}_{CR}$ denote the vector obtained from $\widetilde{\boldsymbol{C}}$ by taking the elements associated with the non-reserved subcarriers. Thus the new problem is:

$$\begin{aligned} &\min_{\widetilde{\boldsymbol{C}}} \widetilde{\boldsymbol{C}}^T \boldsymbol{N} \widetilde{\boldsymbol{C}} + \boldsymbol{u}^T \widetilde{\boldsymbol{C}}, \\ &\text{such that} \quad -\operatorname{diag}\left\{\widetilde{\boldsymbol{X}}_{CR}\right\} \widetilde{\boldsymbol{C}}_{CR} \preceq \boldsymbol{0}, \\ &\qquad\qquad\quad \boldsymbol{E}_{CR} \widetilde{\boldsymbol{C}}_{CR} = \boldsymbol{0}, \\ &\qquad\qquad\quad \widetilde{\boldsymbol{C}} \preceq L_{up} \boldsymbol{1}, \\ &\qquad\qquad\quad \widetilde{\boldsymbol{C}} \succeq L_{low} \boldsymbol{1}, \end{aligned} \tag{7.32}$$

where the reserved subcarriers have the flexibility to take any value.

Baseband SI-cancellation stage can also be applied at the receiver to offer further SI suppression. A replica of the received SI is obtained using the known transmitted signal, an estimate of the transmitter nonlinearities and an estimate of the SI channel then the obtained replica is subtracted from the received signal. In practice, the baseband cancellation is only possible if the SI is already attenuated to avoid saturation of the receiver components. Baseband cancellation schemes that suppress the transmitter impairments have been studied in [7, 8] and Chaps. 4–6. These schemes are based on the presumption that the receiver knows its transmitted data and models the transmitter impairments to have an accurate approximation of the actual received SI.

7.5 Simulation Results

In this section, we evaluate and compare the performance of the proposed cancellation schemes applied to a 4 × 4 MIMO full-duplex system using OFDM-4QAM with $N = 128$ subcarriers. The wireless channels are represented by multipath Rayleigh fading with five paths. A complete transmission chain is implemented to model the PA, IQ mixer, LNA and ADC responses. The PA coefficients are derived from the intercept points by taking IIP3 = 20 dBm. The image rejection ratio of the IQ mixer is set to 28 dB, which is specified in 3GPP LTE specifications [13]. The ADC is realized by a 12-bit uniform quantizer to incorporate the quantization noise. Thus no simplifications are made in the simulations regarding the different impairments. Before cancellation, the SI is assumed to be 100 dB higher than the intended signal. The amount of passive cancellation is 40 dB [9, 14]. Let α_{TR}, α_{CR} and $\alpha_{TR,CR}$ represent the amounts of SI reduction provided by the tone reservation method, the constellation relaxation method and the combination of both methods, respectively. In the following, we evaluate the performance of the cancellation schemes in term of the cumulative distribution function of the SI-cancellation and its average.

We first focus on the case of perfect SI channel knowledge. Figure 7.4 illustrates the performance of the tone reservation method. The plots show the cumulative distribution function of α_{TR} for different numbers of reserved subcarriers. In Fig. 7.4, we vary the number of reserved subcarriers used to reduce the SI from 25% to 12.5% to study the SI reduction versus the data rate trade-off. As expected, more reserved subcarriers provide better SI-reduction performance at the costs of lowering the data transmission rate. An average SI-reduction of 17.73 dB and 13 dB are obtained with 25% and 12.5% reserved subcarriers, respectively. The choice of the reserved subcarriers is an essential part of the tone reservation method. For the same number of reserved subcarriers using exhaustive search, we find the optimal reserved subcarriers placement (for the highest SI reduction). As shown in

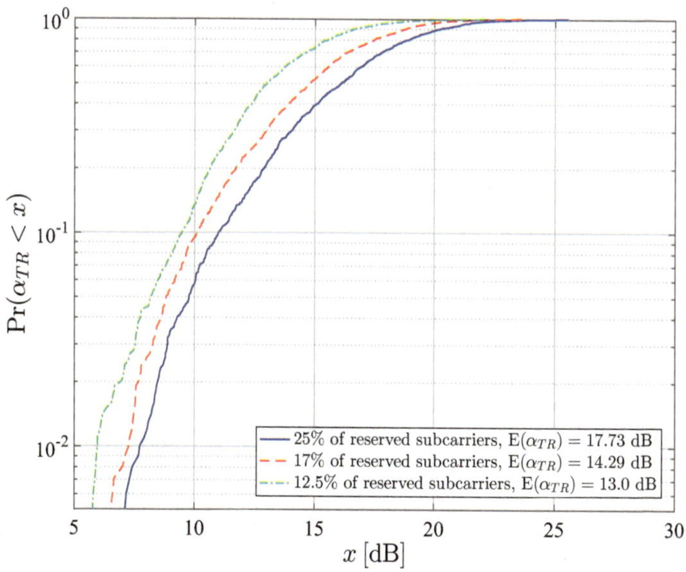

Fig. 7.4 SI-reduction performance of the tone reservation scheme when $N_t = N_r = 4$, and the SI channel is perfectly known

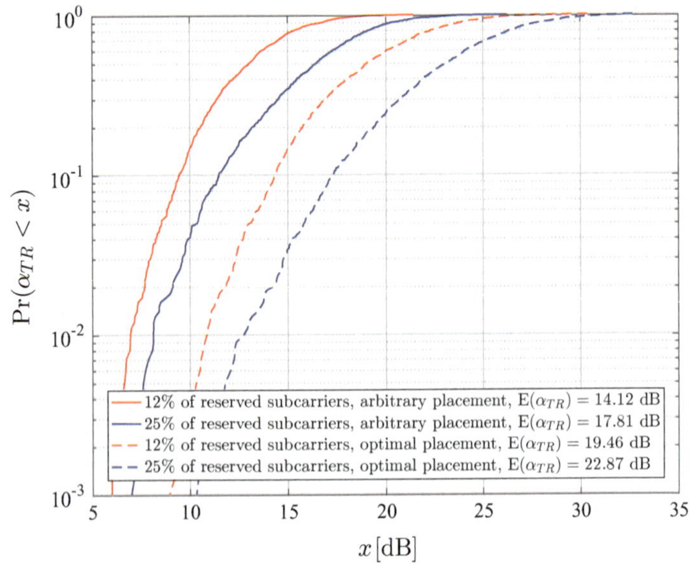

Fig. 7.5 SI-reduction performance of the tone reservation method with optimal and arbitrary reserved-subcarrier placement

Fig. 7.5, the optimal placement, using exhaustive search, improves the SI-reduction performance by about 6 dB. However, the complexity of the exhaustive search increases with the number of subcarriers and quickly becomes intractable. In the following results, we arbitrary select the reserved subcarriers.

To justify the effectiveness of the problem formulation in (7.7), we evaluate through simulations the SI reduction obtained at each antenna, which is essentially the same for different antennas as shown in Fig. 7.6. Thus, we avoid the saturation of all LNA/ADC in the different antennas.

As explained in Sect. 7.3, the extended points result in increasing the average power, which can be controlled by adjusting the lower and upper bounds on the values of the extended points as given in (7.30). Figure 7.7 plots the amount of SI reduction when using the constellation relaxation method. As the limits on the extended points increase, there is more space that can be used to design the cancelling signal which provides a larger SI reduction. The plots show that a reduction of 17.48 dB is obtained on average with a small increase of 1.4 dB in the transmit power. The "power increase" in Fig. 7.7 represents the maximum allowable power increase, specified by the constraint in (7.30). We evaluate through simulations the actual power increase and the resulting amount of reduction when fixing the maximum power increase, where the results are reported in Table 7.1.

For 4-QAM, all symbols can be extended since they are located at the corners of the constellation. For higher-order constellations, the interior points are not extended, which reduces the flexibility to design the cancelling signal. To study the performance of the proposed method for higher-order constellations, Table 7.2

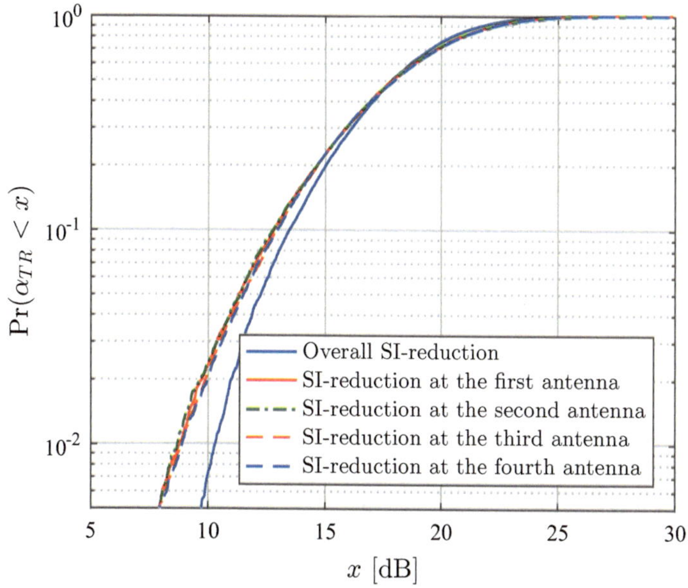

Fig. 7.6 SI-reduction performance of the tone reservation method at each antenna separately

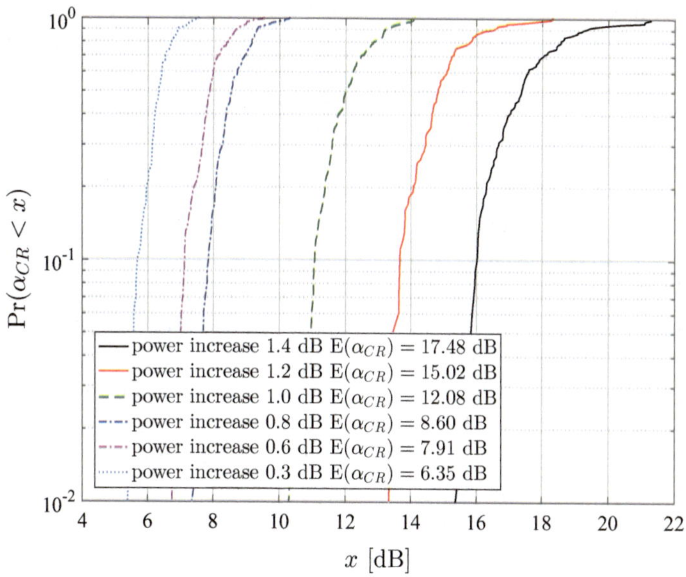

Fig. 7.7 SI-reduction performance of the constellation relaxation scheme when $N_t = N_r = 4$, and the SI channel is perfectly known

Table 7.1 Power increase of the constellation relaxation method

Maximum power increase in dB	Actual power increase in dB	Average amount of reduction in dB
0.3	0.26	6.35
0.6	0.42	7.91
0.8	0.67	8.6
1	0.91	12.08
1.2	1.17	15.02
1.4	1.26	17.48

Table 7.2 Amount of reduction for different M-QAM order

Modulation	$E(\alpha_{CR})$ with 1.4 dB power increase	$E(\alpha_{CR})$ with 1.2 dB power increase
4-QAM	17.48 dB	15.02 dB
16-QAM	14.86 dB	12.04 dB
64-QAM	11.39 dB	9.39 dB
256-QAM	10.16 dB	8.52 dB

presents the amount of reduction obtained for different modulation orders. It can be seen that the amount of reduction is reduced when increasing the modulation order.

By combining the two proposed methods, it is possible to avoid their drawbacks while obtaining good cancellation results. Figure 7.8 illustrates the cancellation performance of the combined schemes. With the chosen parameters, the number of the reserved subcarriers is reduced by three times as compared to using only the tone reservation method, while obtaining the same SI-reduction. Actually, combining the two methods offers more flexibility to design the cancelling signal which offers higher SI-reduction, which can go up to 36 dB on average.

Table 7.3 summarizes the SI-reduction performance of the various RF SI cancellation stages in [9, 10, 15], and the proposed ASI schemes using tone reservation, constellation relaxation and combined method. The results shows that the proposed ASI schemes can offer competitive performance without additional hardware complexity and power consumption for SI-replica estimation, construction and subtraction needed in the RF SI cancellation stage.

Now we consider the practical case of the imperfect SI channel. An estimate of the SI channel is obtained during a half-duplex period where each transmitter receives only its own signal. Therefore, the SI channel can be estimated using the simple least-square estimator. Figure 7.9 illustrates the SI-reduction performance of different reduction schemes with an estimated SI channel. We compare the proposed ASI methods with the spatial cancellation methods in [1]. Note that spatial cancellation methods reduce the input and output dimensions by using N_t^{SC} and N_r^{SC} transmit and receive streams for data transmission. We consider for comparison the antenna selection (AS) and the beam selection (BS) methods [1] where the BS is equivalent to the null-space projection if N_t^{SC}, N_r^{SC} and the rank of the SI channel are low with respect to N_t and N_r. Both antenna selection and beam

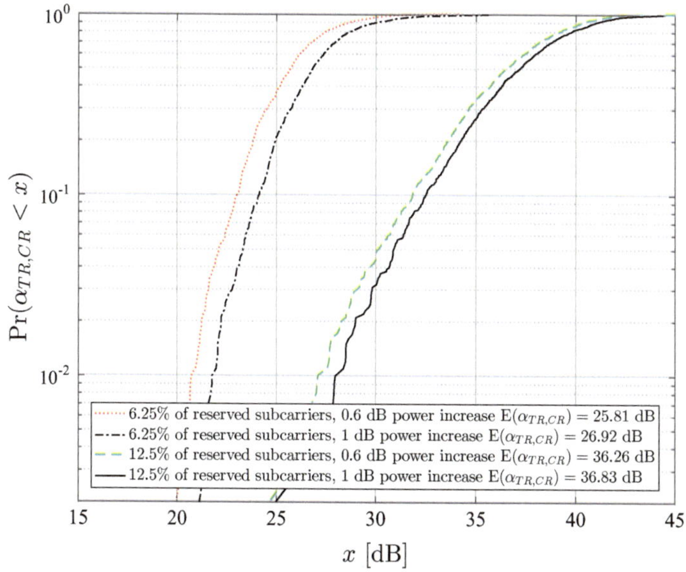

Fig. 7.8 SI-reduction performance of the combined methods when $N_t = N_r = 4$, and the SI channel is perfectly known

Table 7.3 Performance comparison of the RF SI cancellation and the ASI techniques

Technique:	RF SI cancellation [10]	RF SI cancellation [9]	RF SI cancellation [15]	ASI tone reservation	ASI constellation relaxation	ASI combined method
Average SI-reduction (dB)	30	13	40	17	17	36
Bandwidth (MHz)	0.625	15	20	20	20	20

selection are sensitive to input and output stream $N_t^{SC} \times N_r^{SC}$. α_{BS} and α_{AS} refer to the reduction obtained by the BS and AS methods, respectively. Figure 7.9 shows that the spatial cancellation outperforms the tone reservation and the constellation relaxation if the beam selection is used with $N_t^{SC} = N_r^{SC} = 2$ transmit and receive streams. On the other hand, the tone reservation and the constellation relaxation techniques offer significantly higher spectral efficiency by transmitting four independent streams. Moreover, combining the two proposed techniques offers better SI-reduction performance than the spatial cancellation methods, while using all antennas for data transmission.

Figure 7.10 compares the proposed ASI methods with the spatial cancellation in term of bit error rate (BER) versus the initial signal-to-interference power ratio (SIR) without any cancellation. The SNR is fixed at 20 dB. Here, an adaptive baseband cancellation stage is applied at the receiver by estimating and subtracting

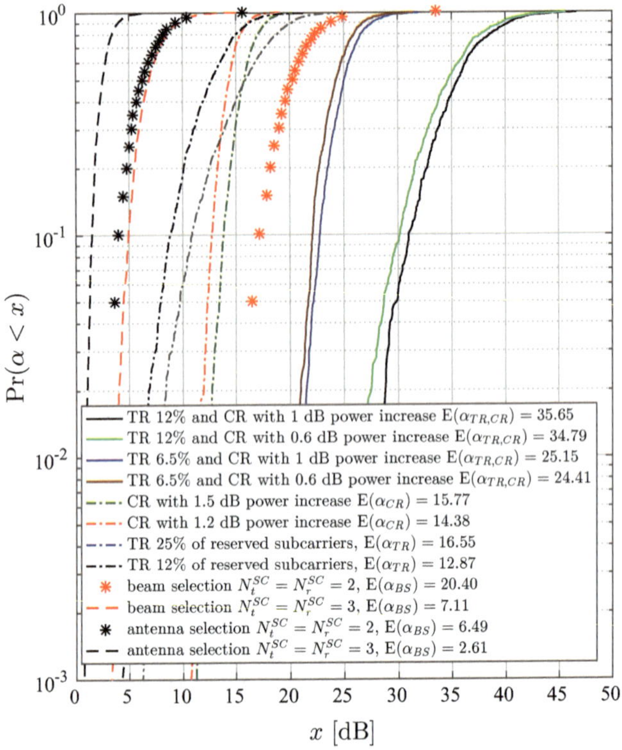

Fig. 7.9 SI-reduction performance of the proposed cancellation methods with $N_t = N_r = 4$ and estimated SI channel

the residual SI. Then, we use linear zero-forcing decoder to detect the intended data from the other transmitter. If the SI is sufficiently reduced or with large initial SIR, the full-duplex system achieves almost the same BER at a half-duplex system. One counter-intuitive feature from Fig. 7.10 is that the AS and BS methods offer higher BER (i.e., worse performance) than the proposed ASI methods, even when the AS and BS provide higher SI-reduction than the tone reservation or constellation relaxation methods. Actually, when implementing the AS or the BS, both transmit and receive filters are applied on the transmitted and received signals, respectively. The SI reduction shown in Fig. 7.9 is measured at the output of the receive filter. However, the receive filter is performed at the baseband, after the ADC, which makes the quantization noise very large and cannot be reduced by the baseband cancellation stage. Whereas the proposed ASI methods reduce the SI before the ADC, making additional baseband processing possible [16].

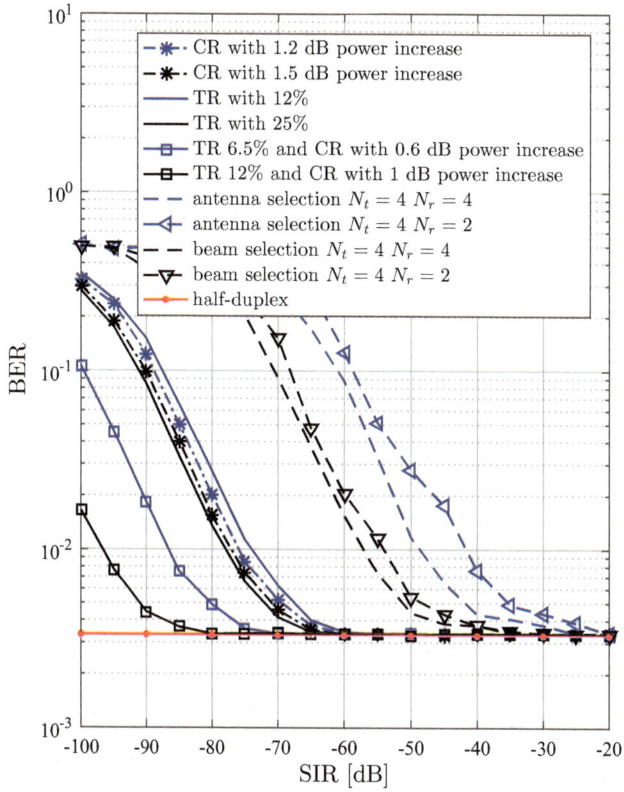

Fig. 7.10 BER vs SIR for the difference cancellation techniques with $N_t = N_r = 4$ for the ASI, $N_t^{SC} = N_r^{SC} = 2$, and estimated SI channel

7.6 Chapter Summary

This chapter presented a new active signal injection (ASI) scheme to reduce the SI in full-duplex systems. The proposed scheme reduces the SI by designing a cancelling signal that is transmitted with the data-bearing signal. The ASI scheme allows for different techniques to be developed. The main criteria when developing such methods are to reduce the SI at the receiver input and the data detection should not be affected by the cancelling signal. In particular, we proposed the tone reservation technique by sending the cancelling signal on some dedicated subcarriers of the OFDM signal and the constellation relaxation technique by allowing the transmitted data to take one value from a whole range. Compared to the RF SI cancellation, the proposed ASI approach shares the same objective of suppressing SI prior to LNA/ADC to avoid overloading/saturation without additional complexity and power consumption for SI-replica estimation, construction and subtraction needed in the RF SI cancellation stage. The proposed ASI approach can be classified as a spatial SI cancellation scheme that does not reduce the input and output dimensions of the available MIMO system. A marginal number of subcarriers are used to design the injected signal and/or a small increase of about 1 dB in the transmit power represent the main drawback of the proposed ASI.

Appendix 1: Development of the Cost Function for the Tone Reservation Method

By defining $\boldsymbol{C}_{\mathscr{Q}_k^c} = \left[C_{q_1}(N-k), \ldots, C_{q_n}(N-k)\right]^T$ and $\widetilde{\boldsymbol{d}}_{r,\mathscr{Q}_k^c} = \left[g_{2,q_1} d_{rq_1}(k), \ldots, \right]$ $\left[g_{2,q_n} d_{rq_n}(k)\right]^T$ for $\{q_1, \ldots, q_n\} \in \mathscr{Q}_{N-k}$, and following the same strategy as done for $k = 0$ and $k = N/2$ in Sect. 7.2, $|Y_r^{\mathrm{SI}}(k)|^2$ is reformulated as:

$$\begin{aligned} |Y_r^{\mathrm{SI}}(k)|^2 &= \boldsymbol{C}_{\mathscr{Q}_k}^T \boldsymbol{M}_{d,\widetilde{d^*}}^k \boldsymbol{C}_{\mathscr{Q}_k^c} + \boldsymbol{C}_{\mathscr{Q}_k}^T \boldsymbol{M}_{d,d^*}^k \boldsymbol{C}_{\mathscr{Q}_k}^* \\ &+ \boldsymbol{C}_{\mathscr{Q}_k}^H \boldsymbol{M}_{d^*,\widetilde{d}} \boldsymbol{C}_{\mathscr{Q}_k^c}^* + \boldsymbol{C}_{\mathscr{Q}_k^c}^H \boldsymbol{M}_{\widetilde{d},\widetilde{d^*}}^k \boldsymbol{C}_{\mathscr{Q}_k^c} \\ &+ \boldsymbol{d}_{r,\mathscr{Q}_k}^T \boldsymbol{C}_{\mathscr{Q}_k} Z_{r,k}^* + \widetilde{\boldsymbol{d}}_{r,\mathscr{Q}_k^c}^T \boldsymbol{C}_{\mathscr{Q}_k^c}^* Z_{r,k}^* \\ &+ Z_{r,k} \left(\boldsymbol{d}_{r,\mathscr{Q}_k}^H \boldsymbol{C}_{\mathscr{Q}_k}^* + \widetilde{\boldsymbol{d}}_{r,\mathscr{Q}_k^c}^H \boldsymbol{C}_{\mathscr{Q}_k^c} + Z_{r,k}^* \right). \end{aligned} \tag{7.33}$$

In (7.33), $\boldsymbol{M}_{d,\widetilde{d^*}}^k = \boldsymbol{d}_{r,\mathscr{Q}_k} \widetilde{\boldsymbol{d}}_{r,\mathscr{Q}_k^c}^H$, $\boldsymbol{M}_{d^*,\widetilde{d}}^k = \boldsymbol{d}_{r,\mathscr{Q}_k}^* \widetilde{\boldsymbol{d}}_{r,\mathscr{Q}_k^c}^T$ and $\boldsymbol{M}_{\widetilde{d},\widetilde{d^*}}^k = \widetilde{\boldsymbol{d}}_{r,\mathscr{Q}_k^c} \widetilde{\boldsymbol{d}}_{r,\mathscr{Q}_k^c}^H$. By expending $\boldsymbol{C}_{\mathscr{Q}_k}$ and $\boldsymbol{C}_{\mathscr{Q}_k^c}$ to their real and imaginary parts as $\widetilde{\boldsymbol{C}}_{\mathscr{Q}_k} = \left[\Re\{\boldsymbol{C}_{\mathscr{Q}_k}^T\}\ \Im\{\boldsymbol{C}_{\mathscr{Q}_k}^T\}\right]^T$ and $\widetilde{\boldsymbol{C}}_{\mathscr{Q}_k^c} = \left[\Re\{\boldsymbol{C}_{\mathscr{Q}_k^c}^T\}\ \Im\{\boldsymbol{C}_{\mathscr{Q}_k^c}^T\}\right]^T$, (7.33) is rewritten as:

$$\begin{aligned} |Y_r^{\mathrm{SI}}(k)|^2 = \widetilde{\boldsymbol{C}}_{\mathscr{Q}_k}^T &\left(\begin{pmatrix} \boldsymbol{M}_{d,\widetilde{d^*}}^k & j\boldsymbol{M}_{d,\widetilde{d^*}}^k \\ j\boldsymbol{M}_{d,\widetilde{d^*}}^k & -\boldsymbol{M}_{d,\widetilde{d^*}}^k \end{pmatrix} \widetilde{\boldsymbol{C}}_{\mathscr{Q}_k^c} + \begin{pmatrix} \boldsymbol{M}_{d,d^*}^k & -j\boldsymbol{M}_{d,d^*}^k \\ j\boldsymbol{M}_{d,d^*}^k & \boldsymbol{M}_{d,d^*}^k \end{pmatrix} \widetilde{\boldsymbol{C}}_{\mathscr{Q}_k} \right) \\ + \widetilde{\boldsymbol{C}}_{\mathscr{Q}_k^c}^T &\left(\begin{pmatrix} \boldsymbol{M}_{d^*,\widetilde{d}}^k & -j\boldsymbol{M}_{d^*,\widetilde{d}}^k \\ -j\boldsymbol{M}_{d^*,\widetilde{d}}^k & -\boldsymbol{M}_{d^*,\widetilde{d}}^k \end{pmatrix} \widetilde{\boldsymbol{C}}_{\mathscr{Q}_k} + \begin{pmatrix} \boldsymbol{M}_{\widetilde{d},\widetilde{d^*}}^k & j\boldsymbol{M}_{\widetilde{d},\widetilde{d^*}}^k \\ -j\boldsymbol{M}_{\widetilde{d},\widetilde{d^*}}^k & \boldsymbol{M}_{\widetilde{d},\widetilde{d^*}}^k \end{pmatrix} \widetilde{\boldsymbol{C}}_{\mathscr{Q}_k^c} \right) \\ &+ 2\left[\Re\{\boldsymbol{d}_{r,\mathscr{Q}_k}^T Z_{r,k}^*\}\ \Im\{\boldsymbol{d}_{r,\mathscr{Q}_k}^H Z_{r,k}\}\right] \widetilde{\boldsymbol{C}}_{\mathscr{Q}_k} \\ &+ 2\left[\Re\{\widetilde{\boldsymbol{d}}_{r,\mathscr{Q}_k^c}^T Z_{r,k}^*\}\ \Im\{\widetilde{\boldsymbol{d}}_{r,\mathscr{Q}_k^c}^T Z_{r,k}^*\}\right] \widetilde{\boldsymbol{C}}_{\mathscr{Q}_k^c} + |Z_{r,k}|^2. \end{aligned} \tag{7.34}$$

On the other hand, $|Y_r^{\mathrm{SI}}(N-k)|^2$ can also be written as:

$$|Y_r^{\mathrm{SI}}(N-k)|^2 = \widetilde{\boldsymbol{C}}_{\mathscr{Q}_k}^T \left(\begin{pmatrix} \boldsymbol{M}_{\widetilde{d,d^*}}^{N-k} & -j\boldsymbol{M}_{\widetilde{d,d^*}}^{N-k} \\ -j\boldsymbol{M}_{\widetilde{d,d^*}}^{N-k} & -\boldsymbol{M}_{\widetilde{d,d^*}}^{N-k} \end{pmatrix} \widetilde{\boldsymbol{C}}_{\mathscr{Q}_k^c} + \begin{pmatrix} \boldsymbol{M}_{\widetilde{d,d^*}}^{N-k} & j\boldsymbol{M}_{\widetilde{d,d^*}}^{N-k} \\ -j\boldsymbol{M}_{\widetilde{d,d^*}}^{N-k} & \boldsymbol{M}_{\widetilde{d,d^*}}^{N-k} \end{pmatrix} \widetilde{\boldsymbol{C}}_{\mathscr{Q}_k} \right)$$

$$+ \widetilde{\boldsymbol{C}}^T_{\mathscr{Q}_k^c} \left(\begin{pmatrix} \boldsymbol{M}^{N-k}_{\widetilde{d,d^*}} & j\boldsymbol{M}^{N-k}_{\widetilde{d,d^*}} \\ j\boldsymbol{M}^{N-k}_{\widetilde{d,d^*}} & -\boldsymbol{M}^{N-k}_{\widetilde{d,d^*}} \end{pmatrix} \widetilde{\boldsymbol{C}}_{\mathscr{Q}_k} + \begin{pmatrix} \boldsymbol{M}^{N-k}_{d,d^*} & -j\boldsymbol{M}^{N-k}_{d,d^*} \\ j\boldsymbol{M}^{N-k}_{d,d^*} & \boldsymbol{M}^{N-k}_{d,d^*} \end{pmatrix} \widetilde{\boldsymbol{C}}_{\mathscr{Q}_k^c} \right)$$
$$+ 2\left[\Re\{\widetilde{\boldsymbol{d}}^T_{r,\mathscr{Q}^c_{N-k}} Z^*_{r,N-k}\} \; \Im\{\widetilde{\boldsymbol{d}}^T_{r,\mathscr{Q}^c_{N-k}} Z^*_{r,N-k}\}\right] \widetilde{\boldsymbol{C}}_{\mathscr{Q}_k}$$
$$+ 2\left[\Re\{\boldsymbol{d}^T_{r,\mathscr{Q}_{N-k}} Z^*_{r,N-k}\} \; \Im\{\boldsymbol{d}^H_{r,\mathscr{Q}_{N-k}} Z_{r,N-k}\}\right] \widetilde{\boldsymbol{C}}_{\mathscr{Q}_k^c} + |Z_{r,N-k}|^2. \quad (7.35)$$

Because of the image signal from the IQ mixer, $C_q(k)$ appears at subcarriers k and $N-k$. Therefore, the two expressions in (7.34) and (7.35) are combined to result to the following expression:

$$|Y_r^{\mathrm{SI}}(k)|^2 + |Y_r^{\mathrm{SI}}(N-k)|^2 = \widetilde{\boldsymbol{C}}^T_{\mathscr{Q}_k} \boldsymbol{M}^k_{\mathscr{Q}_k,\mathscr{Q}_k^c} \widetilde{\boldsymbol{C}}_{\mathscr{Q}_k^c} + \widetilde{\boldsymbol{C}}^T_{\mathscr{Q}_k} \boldsymbol{M}^k_{\mathscr{Q}_k,\mathscr{Q}_k} \widetilde{\boldsymbol{C}}_{\mathscr{Q}_k} + \widetilde{\boldsymbol{C}}^T_{\mathscr{Q}_k^c} \boldsymbol{M}^k_{\mathscr{Q}_k^c,\mathscr{Q}_k} \widetilde{\boldsymbol{C}}_{\mathscr{Q}_k}$$
$$+ \widetilde{\boldsymbol{C}}^T_{\mathscr{Q}_k^c} \boldsymbol{M}^k_{\mathscr{Q}_k^c,\mathscr{Q}_k^c} \widetilde{\boldsymbol{C}}_{\mathscr{Q}_k^c} + \boldsymbol{v}^T_{\mathscr{Q}_k} \widetilde{\boldsymbol{C}}_{\mathscr{Q}_k} + \boldsymbol{v}^T_{\mathscr{Q}_k^c} \widetilde{\boldsymbol{C}}_{\mathscr{Q}_k^c} + |Z_{r,k}|^2 + |Z_{r,N-k}|^2, \quad (7.36)$$

or in the following more compact form:

$$|Y_r^{\mathrm{SI}}(k)|^2 + |Y_r^{\mathrm{SI}}(N-k)|^2 = \underbrace{\left[\widetilde{\boldsymbol{C}}^T_{\mathscr{Q}_k} \; \widetilde{\boldsymbol{C}}^T_{\mathscr{Q}_k^c}\right]}_{\widetilde{\boldsymbol{C}}_k^T} \underbrace{\begin{pmatrix} \boldsymbol{M}^k_{\mathscr{Q}_k,\mathscr{Q}_k} & \boldsymbol{M}^k_{\mathscr{Q}_k,\mathscr{Q}_k^c} \\ \boldsymbol{M}^k_{\mathscr{Q}_k^c,\mathscr{Q}_k} & \boldsymbol{M}^k_{\mathscr{Q}_k^c,\mathscr{Q}_k^c} \end{pmatrix}}_{\boldsymbol{M}_{r,k}} \underbrace{\begin{bmatrix} \widetilde{\boldsymbol{C}}_{\mathscr{Q}_k} \\ \widetilde{\boldsymbol{C}}_{\mathscr{Q}_k^c} \end{bmatrix}}_{\widetilde{\boldsymbol{C}}_k}$$
$$+ \underbrace{\left[\boldsymbol{v}^T_{\mathscr{Q}_k} \; \boldsymbol{v}^T_{\mathscr{Q}_k^c}\right]}_{\boldsymbol{v}^T_{r,k}} \begin{bmatrix} \widetilde{\boldsymbol{C}}_{\mathscr{Q}_k} \\ \widetilde{\boldsymbol{C}}_{\mathscr{Q}_k^c} \end{bmatrix} + |Z_{r,k}|^2 + |Z_{r,N-k}|^2$$
$$= \widetilde{\boldsymbol{C}}_k^T \boldsymbol{M}_{r,k} \widetilde{\boldsymbol{C}}_k + \boldsymbol{v}^T_{r,k} \widetilde{\boldsymbol{C}}_k + |Z_{r,k}|^2 + |Z_{r,N-k}|^2, \quad (7.37)$$

where the matrix $\boldsymbol{M}^k_{\mathscr{Q}_k,\mathscr{Q}_k^c}$ collects the common terms between $\widetilde{\boldsymbol{C}}^T_{\mathscr{Q}_k}$ and $\widetilde{\boldsymbol{C}}_{\mathscr{Q}_k^c}$ as:

$$\boldsymbol{M}^k_{\mathscr{Q}_k,\mathscr{Q}_k^c} = \begin{pmatrix} \boldsymbol{M}^k_{\widetilde{d,d^*}} & j\boldsymbol{M}^k_{\widetilde{d,d^*}} \\ j\boldsymbol{M}^k_{\widetilde{d,d^*}} & -\boldsymbol{M}^k_{\widetilde{d,d^*}} \end{pmatrix} + \begin{pmatrix} \boldsymbol{M}^{N-k}_{\widetilde{d,d^*}} & -j\boldsymbol{M}^{N-k}_{\widetilde{d,d^*}} \\ -j\boldsymbol{M}^{N-k}_{\widetilde{d,d^*}} & -\boldsymbol{M}^{N-k}_{\widetilde{d,d^*}} \end{pmatrix}, \quad (7.38)$$

and the other matrices $\boldsymbol{M}^k_{\mathscr{Q}_k,\mathscr{Q}_k}$, $\boldsymbol{M}^k_{\mathscr{Q}_k^c,\mathscr{Q}_k}$ and $\boldsymbol{M}^k_{\mathscr{Q}_k^c,\mathscr{Q}_k^c}$ follow the same principle as:

$$\boldsymbol{M}^k_{\mathscr{Q}_k,\mathscr{Q}_k} = \begin{pmatrix} \boldsymbol{M}^k_{d,d^*} & -j\boldsymbol{M}^k_{d,d^*} \\ j\boldsymbol{M}^k_{d,d^*} & \boldsymbol{M}^k_{d,d^*} \end{pmatrix} + \begin{pmatrix} \boldsymbol{M}^{N-k}_{\widetilde{d,d^*}} & j\boldsymbol{M}^{N-k}_{\widetilde{d,d^*}} \\ -j\boldsymbol{M}^{N-k}_{\widetilde{d,d^*}} & \boldsymbol{M}^{N-k}_{\widetilde{d,d^*}} \end{pmatrix}$$
$$\boldsymbol{M}^k_{\mathscr{Q}_k^c,\mathscr{Q}_k} = \begin{pmatrix} \boldsymbol{M}^k_{d^*,\widetilde{d}} & -j\boldsymbol{M}^k_{d^*,\widetilde{d}} \\ -j\boldsymbol{M}^k_{d^*,\widetilde{d}} & -\boldsymbol{M}^k_{d^*,\widetilde{d}} \end{pmatrix} + \begin{pmatrix} \boldsymbol{M}^{N-k}_{\widetilde{d,d^*}} & j\boldsymbol{M}^{N-k}_{\widetilde{d,d^*}} \\ j\boldsymbol{M}^{N-k}_{\widetilde{d,d^*}} & -\boldsymbol{M}^{N-k}_{\widetilde{d,d^*}} \end{pmatrix}$$
$$\boldsymbol{M}^k_{\mathscr{Q}_k^c,\mathscr{Q}_k^c} = \begin{pmatrix} \boldsymbol{M}^k_{\widetilde{d^*,d}} & j\boldsymbol{M}^k_{\widetilde{d^*,d}} \\ -j\boldsymbol{M}^k_{\widetilde{d^*,d}} & \boldsymbol{M}^k_{\widetilde{d^*,d}} \end{pmatrix} + \begin{pmatrix} \boldsymbol{M}^{N-k}_{d,d^*} & -j\boldsymbol{M}^{N-k}_{d,d^*} \\ j\boldsymbol{M}^{N-k}_{d,d^*} & \boldsymbol{M}^{N-k}_{d,d^*} \end{pmatrix}. \quad (7.39)$$

Also, the vectors $\boldsymbol{v}_{\mathscr{Q}_k}$ and $\boldsymbol{v}_{\mathscr{Q}_k^c}$ are the sum of the coefficients multiplying $\widetilde{\boldsymbol{C}}_{\mathscr{Q}_k}$ and $\widetilde{\boldsymbol{C}}_{\mathscr{Q}_k^c}$, respectively, in (7.34) and (7.35) and given by:

$$\boldsymbol{v}_{\mathscr{Q}_k}=2\left[\Re\{\boldsymbol{d}^T_{r,\mathscr{Q}_k}Z^*_{r,k}\}\ \Im\{\boldsymbol{d}^H_{r,\mathscr{Q}_k}Z_{r,k}\}\right]+2\left[\Re\{\widetilde{\boldsymbol{d}}^T_{r,\mathscr{Q}^c_{N-k}}Z^*_{r,N-k}\}\ \Im\{\widetilde{\boldsymbol{d}}^T_{r,\mathscr{Q}^c_{N-k}}Z^*_{r,N-k}\}\right]$$
$$\boldsymbol{v}_{\mathscr{Q}_k^c}=2\left[\Re\{\widetilde{\boldsymbol{d}}^T_{r,\mathscr{Q}_k^c}Z^*_{r,k}\}\ \Im\{\widetilde{\boldsymbol{d}}^T_{r,\mathscr{Q}_k^c}Z^*_{r,k}\}\right]+2\left[\Re\{\boldsymbol{d}^T_{r,\mathscr{Q}_{N-k}}Z^*_{r,N-k}\}\ \Im\{\boldsymbol{d}^H_{r,\mathscr{Q}_{N-k}}Z_{r,N-k}\}\right]. \tag{7.40}$$

Gathering all the expressions in (7.16) and (7.37), we obtain:

$$||\boldsymbol{y}_r^{\mathrm{SI}}||_2^2=\widetilde{\boldsymbol{C}}^T\boldsymbol{M}_r\widetilde{\boldsymbol{C}}+\boldsymbol{v}_r^T\widetilde{\boldsymbol{C}}+\sum_{k=0}^{N-1}|Z_{r,k}|^2, \tag{7.41}$$

where $\widetilde{\boldsymbol{C}}=\left[\widetilde{\boldsymbol{C}}^T_{\mathscr{Q}_0},\widetilde{\boldsymbol{C}}_1,\ldots,\widetilde{\boldsymbol{C}}_{\frac{N}{2}-1},\widetilde{\boldsymbol{C}}^T_{\mathscr{Q}_{\frac{N}{2}}}\right]^T$, $\boldsymbol{v}_r=\left[\boldsymbol{v}^T_{r,0},\ldots,\boldsymbol{v}^T_{r,\frac{N}{2}}\right]^T$ and $\boldsymbol{M}_r$ is a block diagonal matrix given by:

$$\boldsymbol{M}_r=\begin{pmatrix}\boldsymbol{M}_{r,0} & \boldsymbol{0} & \ldots & \boldsymbol{0}\\ \boldsymbol{0} & \boldsymbol{M}_{r,1} & \ldots & \boldsymbol{0}\\ \vdots & & \ddots & \vdots\\ \boldsymbol{0} & \ldots & \boldsymbol{0} & \boldsymbol{M}_{r,\frac{N}{2}}\end{pmatrix}. \tag{7.42}$$

It follows that the cost function in (7.7) is written as:

$$\sum_{r=1}^{N_r}||\boldsymbol{y}_r^{\mathrm{SI}}||_2^2=\widetilde{\boldsymbol{C}}^T\underbrace{\sum_{r=1}^{N_r}\boldsymbol{M}_r}_{\boldsymbol{M}}\widetilde{\boldsymbol{C}}+\underbrace{\sum_{r=1}^{N_r}\boldsymbol{v}_r^T}_{\boldsymbol{v}^T}\widetilde{\boldsymbol{C}}+\sum_{r=1}^{N_r}\sum_{k=0}^{N-1}|Z_{r,k}|^2, \tag{7.43}$$

which gives the result in (7.18).

Appendix 2: On the Invertibility of the Matrix *M*

From (7.42) and (7.43), the matrix $\boldsymbol{M}$ is a block diagonal matrix. Therefore, to show that $\boldsymbol{M}$ is invertible, it suffices to show that every block $\sum_{r=1}^{N_r}\boldsymbol{M}_{r,k}$, for $k=0,\ldots,N/2$, is full-rank. For $k=0$ and $k=N/2$, $\boldsymbol{M}_{r,k}$ is the sum of 4 square matrices, each of rank 1 with size $2\mathrm{card}(\mathscr{Q}_k)$, where $\mathrm{card}(\mathscr{Q}_k)$ denotes the number of elements in $\mathscr{Q}_k$. Therefore, $\boldsymbol{M}_{r,k}$ is of rank $\min(4,\ 2\mathrm{card}(\mathscr{Q}_k))$. It follows that, if $4N_t<2\mathrm{card}(\mathscr{Q}_k)$, the matrix $\sum_{r=1}^{N_r}\boldsymbol{M}_{r,k}$ is of rank $4N_t$ otherwise it is of rank $2\mathrm{card}(\mathscr{Q}_k)$ (i.e., full-rank). When choosing the reserved subcarriers, we allocate one antenna to every subcarrier, so that $\text{card}(\mathcal{Q}_k) = 1$ (most of the time), or 2, depending on the total number of reserved subcarriers. As a result, we have $4N_t > 2\text{card}(\mathcal{Q}_k)$ and the matrix $\sum_{r=1}^{N_r} \boldsymbol{M}_{r,k}$ is full-rank, for $k = 0$ and $N/2$. By the same reasoning, we can prove that the other matrices, $\sum_{r=1}^{N_r} \boldsymbol{M}_{r,k}$ are also full-rank, for $k = 1, \ldots, N/2 - 1$.

References

1. T. Riihonen, S. Werner, and R. Wichman, "Mitigation of loopback self-interference in full-duplex MIMO relays," *IEEE Trans. Signal Process.*, vol. 59, no. 12, pp. 5983–5993, 2011.
2. Y. Hua, P. Liang, Y. Ma, A. C. Cirik, and Q. Gao, "A method for broadband full-duplex MIMO radio," *IEEE Signal Process. Letters*, vol. 19, no. 12, pp. 793–796, 2012.
3. B. Chun, E.-R. Jeong, J. Joung, Y. Oh, and Y. H. Lee, "Pre-nulling for self-interference suppression in full-duplex relays," in *Proc. APSIPA ASC 2009: Asia-Pacific Signal and Information Processing Association*, Xi'an, China, Oct. 2009.
4. A. Masmoudi and T. Le-Ngoc, "Self-interference mitigation using active signal injection full-duplex MIMO-OFDM systems," in *Proc. IEEE Veh. Technol. Conf.*, Montréal, Quebec, Canada, Sept. 2016.
5. J. Tellado and J. M. Cioffi, "Peak power reduction for multicarrier transmission," in *Proc. IEEE Global Telecommun. Conf.*, Sydney, Australia, Dec. 1998.
6. B. S. Krongold and D. L. Jones, "PAR reduction in OFDM via active constellation extension," *IEEE Trans. on Broadcasting*, vol. 49, no. 3, pp. 258–268, 2003.
7. A. Masmoudi and T. Le-Ngoc, "A maximum-likelihood channel estimator for self-interference cancellation in full-duplex systems," *IEEE Trans. Vehicular Technology*, vol. 65, no. 7, pp. 5122–5132, July 2016.
8. ——, "Channel estimation and self-interference cancellation in full-duplex communication systems," *IEEE Trans. Vehicular Technology*, vol. 66, no. 1, pp. 321–334, Jan. 2017.
9. J. I. Choi, M. Jain, K. Srinivasan, P. Levis, and S. Katti, "Achieving single channel, full duplex wireless communication," in *Proc. ACM MobiCom*, Chicago, Illinois, USA, Sept. 2010.
10. M. Duarte, C. Dick, and A. Sabharwal, "Experiment-driven characterization of full-duplex wireless systems," *IEEE Trans. Wireless Comm.*, vol. 11, no. 12, pp. 4296–4307, 2012.
11. A. Masmoudi and T. Le-Ngoc, "Self-interference cancellation for full-duplex MIMO transceivers," in *Proc. IEEE Wireless Commun. and Netw. Conf.*, New Orleans, Louisiana, USA, Mar. 2015.
12. S. Boyd and L. Vandenberghe, *Convex optimization*. Cambridge university press, 2004.
13. "LTE; evolved universal terrestrial radio access (E-UTRA); user equipment (UE) radio transmission and reception (3GPP TS 36.101 version 11.2.0 release 11)." ETSI, Sophia Antipolis Cedex, France.
14. E. Everett, A. Sahai, and A. Sabharwal, "Passive self-interference suppression for full-duplex infrastructure nodes," *IEEE Trans. Wireless Comm.*, vol. 13, no. 2, pp. 680–694, 2014.
15. D. Bharadia, E. McMilin, and S. Katti, "Full duplex radios," in *Proc. of the ACM SIGCOMM 2013*, Hong Kong, China, Aug. 2013.
16. A. Masmoudi and T. Le-Ngoc, "Residual self-interference after cancellation in full-duplex systems," in *Proc. IEEE Int. Conf. Commun.*, Sydney, Australia, June 2014.

Chapter 8
Conclusions and Future Work

Abstract Along this book, we consider the problem of SI-cancellation in full-duplex wireless communications systems. Conventional approaches subtract the SI from the received signal and thus reveal the need to estimate and reconstruct the received SI. In the previous works, the proposed estimators ignore the intended signal by either using a training period to estimate the SI parameters, resulting in a reduced spectral efficiency, or simply considering the intended signal as additive noise. In our work, we incorporate the intended signal in the estimation process by exploiting its second-order statistics and the transmitted pilots. We develop new estimation algorithms for SI-cancellation that achieve superior accuracy and spectral efficiency than the available approaches. The main contributions and corresponding results are summarized as follows.

8.1 Concluding Remarks

Along this book, we consider the problem of SI-cancellation in full-duplex wireless communications systems. Conventional approaches subtract the SI from the received signal and thus reveal the need to estimate and reconstruct the received SI. In the previous works, the proposed estimators ignore the intended signal by either using a training period to estimate the SI parameters, resulting in a reduced spectral efficiency, or simply considering the intended signal as additive noise. In our work, we incorporate the intended signal in the estimation process by exploiting its second-order statistics and the transmitted pilots. We develop new estimation algorithms for SI-cancellation that achieve superior accuracy and spectral efficiency than the available approaches. The main contributions and corresponding results are summarized as follows.

The strong SI imposes multiple cancellation stages at the receiver. In Chap. 3, we studied the SI before and after each cancellation stage, taking into account the transceiver impairments. Specific emphasis was put on the reference signal for the RF and baseband cancellation stages. This analysis identified the main limiting factors for SI-cancellation. We clearly justify the need to reduce the SI before the LNA/ADC, using the RF cancellation stage, to avoid high quantization noise from the ADC. The analysis also revealed that the transmitter nonlinearities have to be modeled and cancelled in the baseband cancellation stage, which was treated in the

T. Le-Ngoc, A. Masmoudi, *Full-Duplex Wireless Communications Systems*, Wireless Networks, DOI 10.1007/978-3-319-57690-9_8

following chapters. Once the transmitter nonlinearities were suppressed, the phase noise was identified as the main bottleneck that prevents from completely cancelling the SI.

In Chap. 4, we focused on the problem of parameter estimation for SI-cancellation. A CS-based estimator was proposed to estimate the SI channel for the RF cancellation stage. This estimate is obtained during a short initial half-duplex period. Then, in the presence of the intended signal, we proposed a subspace-based algorithm to estimate the SI channel, the transmitter nonlinearities and the intended channel for the baseband cancellation stage. Including the intended received signal in the estimation process is the main advantage of the proposed algorithm compared to previous works that assume the intended signal as an additive noise. By using the covariance matrix of the received signal, it is possible to obtain the noise subspace without knowing the intended signal. From there, the channel coefficients are obtained up to an ambiguity term which is recovered, along with the transmitter nonlinearities, using the known transmitted SI and the pilot symbols in the intended signal. We used the SINR after cancellation to show that the proposed subspace estimator provides superior accuracy and spectral efficiency than the traditional LS estimator due to its reduced pilot requirements.

In full-duplex, the received signal in the baseband consists of the residual SI and intended signals, the dimension of the signal subspace in full-duplex operation is at least twice than in traditional half-duplex operation. The subspace estimator proposed in the previous chapter requires the number of transmitting antennas to be double of the receiving antennas. In Chap. 5, we circumvented this requirement by processing the received signal and its conjugate. This procedure allows us to double the observation space under some conditions on the pseudo-covariance of the transmitted signal. Also, an iterative procedure is developed to recover the ambiguity term and detect the intended data. Simulations show that the proposed estimator offers better cancellation capability, using one pilot symbol from the intended transmitter, than the LS estimator. Also, an accurate estimate of the intended channel is obtained from one pilot symbol.

Still within the context of parameter estimation and suppression techniques, we explore in Chap. 6 the ML estimation by using both pilots and unknown data from the intended transceiver. To avoid the high complexity of maximizing the true likelihood function, we decoupled the covariance of the intended signal from the intended channel. This resulted to an approximate closed-form solution and allowed us to develop a simple iterative method. We also proposed a phase noise mitigation technique. We first estimate the phase noise affecting the transmitted and received SI then we rotate the reference baseband signal with the estimated phase noise before subtraction.

In summary, in these chapters, we developed various SI channel and nonlinearities estimators suitable for full-duplex wireless communications systems with higher estimation accuracy and better SI-cancellation capability.

Finally, the novel SI-cancellation technique proposed in Chap. 7 represents an alternative to the RF cancellation stage. By transmitting a cancelling signal with the useful signal, the received SI can be considerably reduced. We proposed two methods to design the cancelling signal, namely the tone reservation method and the constellation relaxation method. We obtained a closed-form expression of the cancelling signal when using the tone reservation method, making it easy to implement. Also, the constellation relaxation method can be implemented using quadratic programming. The side effects are a marginal decrease in the effective SNR on the order of 1 dB, and/or a loss of a small number of subcarriers to transmit the cancelling signal. By combining both methods, we avoid their limitation while obtaining a good SI reduction.

8.2 Future Work

Chapters 4–6 showed that including the intended signal in the estimation offers a considerable gain when cancelling the SI. This motivates us to explore other directions related to these techniques.

When developing the ML estimator in Chap. 6, the equivalent channels of the direct signal and the image signal from the IQ mixer, for example, are supposed to be independent. However, they are related by a multiplicative factor that represents the response of the IQ mixer to the image signal. Exploiting this relation can further improve the estimation accuracy. Also, still with the ML estimator, the assumption of independent covariance matrix and intended channel coefficients allowed us to develop the estimator which is an approximation of the ML estimator. However, keeping this dependency will lead to the true ML estimator. One way to solve the problem in this case is to use numerical methods such as Newton-type algorithms.

In Chap. 6, we proposed a method to mitigate the phase noise. For instance, we cannot apply the same approach to estimate the phase noise using a subspace-based algorithm as the phase noise is a time varying process and the resulting subspace will change from one symbol to another. In this case, we can approximate the time-varying process by a basis expansion model, where the problem of estimating the time-varying phase noise reduces to the estimation of a set of static coefficients.

When implementing the tone reservation method proposed in Chap. 7, the reserved subcarriers are chosen in a random manner. A more judicious placement strategy can offer higher SI-cancellation. Also, a rigorous study on the trade-off between SI-cancellation and the transmission rate is certainly useful. This study may reveal an optimal number of reserved subcarriers to sufficiently reduce the SI while maximizing the transmission rate. The same study could be done for the constellation relaxation method where here we optimize the upper and lower bounds that limit the extended points given in (7.30).

The proposed ASI method is well adapted for OFDM-MIMO systems. While the constellation relaxation method can directly be applied to non-OFDM modulated signal, we may need to rethink the tone reservation method. Also, the MIMO system used to transmit different cancellation signals on each antenna such that the combined signal at the receiver is close to zero. Thus the method cannot be directly applied for SISO systems. One way to adapt the method is to consider the signals coming from the different reflection as virtually transmitted by multiple antennas and design the cancelling signal according to this model.

In a cellular network, the base station can operate in full-duplex by transmitting on the downlink to one user and receiving on the uplink from another user simultaneously over the same frequency slot. The two users can operate in half-duplex to avoid SI. But the uplink user will interfere with the downlink user, in particular when the two users are located close to each other, arising another kind of interference. Thus other methods should be applied to manage this interference.